Social Science Perspectives on Climate Change

Although it is generally accepted that the climate is changing for the worse and that human activities are a major contributing factor in that change, there is still only marginal response to the challenge posed by climate change. The reasons behind this limited response are becoming clearer through the recognition that climate change is not just a set of physical science facts, but it is also part of a series of complex social processes. Consequently, this book is important in providing social science perspectives on a range of attempts to adjust human activity to reduce its environmental impact. These attempts vary from the changing of the dress code in Japanese offices to the creation of zero-carbon, gated communities in Bangalore, India. Taken together, the contributions to this book provide timely insights into the complexities of saving the planet through human endeavour.

This book was originally published as a special issue of *Contemporary Social Science*.

David Canter is an applied social psychologist who started his career in the 1960s as an environmental psychologist at Strathclyde University, Glasgow, UK. Since then he has published the results of his research on a wide range of topics, including the design of offices, hospitals and schools, behaviour in fires and emergencies, and complementary medicine. His 1977 book *The Psychology of Place* is still widely cited. Over the last quarter of a century he has developed the discipline of Investigative Psychology. He is currently editor of *Contemporary Social Science,* the journal of the Academy of Social Science.

W0234966

Contemporary Issues in Social Science
Series editor: David Canter
University of Huddersfield, UK

Contemporary Social Science, the journal of the **Academy of Social Sciences**, is an interdisciplinary, cross-national journal which provides a forum for disseminating and enhancing theoretical, empirical and/or pragmatic research across the social sciences and related disciplines. Reflecting the objectives of the Academy of Social Sciences, it emphasises the publication of work that engages with issues of major public interest and concern across the world, and highlights the implications of that work for policy and professional practice.

The *Contemporary Issues in Social Science* book series contains the journal's most cutting-edge special issues. Leading scholars compile thematic collections of articles that are linked to the broad intellectual concerns of *Contemporary Social Science,* and as such these special issues are an important contribution to the work of the journal. The series editor works closely with the guest editor(s) of each special issue to ensure they meet the journal's high standards. The main aim of publishing these special issues as a series of books is to allow a wider audience of both scholars and students from across multiple disciplines to engage with the work of *Contemporary Social Science* and the Academy of Social Sciences.

Social Science Perspectives on Climate Change

Edited by
David Canter

Routledge
Taylor & Francis Group

LONDON AND NEW YORK

ACADEMY
of SOCIAL SCIENCES

First published 2016
by Routledge

4 Park Square, Milton Park, Abingdon, Oxon OX14 4RN
605 Third Avenue, New York, NY 10017

First issued in paperback 2017

Routledge is an imprint of the Taylor & Francis Group, an informa business

British Library Cataloguing in Publication Data
A catalogue record for this book is available from the British Library

Typeset in Times New Roman
by RefineCatch Limited, Bungay, Suffolk

Publisher's Note
The publisher accepts responsibility for any inconsistencies that may have
arisen during the conversion of this book from journal articles to book chapters,
namely the possible inclusion of journal terminology.

Disclaimer
Every effort has been made to contact copyright holders for their permission to
reprint material in this book. The publishers would be grateful to hear from any
copyright holder who is not here acknowledged and will undertake to rectify
any errors or omissions in future editions of this book.

ISBN 13: 978-1-138-09504-5 (pbk)
ISBN 13: 978-1-138-92467-3 (hbk)

Contents

Citation Information

The following chapters were originally published in *Contemporary Social Science*, volume 9, issue 4 (December 2014). When citing this material, please use the original page numbering for each article, as follows:

Chapter 2
Critical issues in social science climate change research
Catherine Leyshon
Contemporary Social Science, volume 9, issue 4 (December 2014) pp. 359–373

Chapter 3
Values, identity and pro-environmental behaviour
Birgitta Gatersleben, Niamh Murtagh and Wokje Abrahamse
Contemporary Social Science, volume 9, issue 4 (December 2014) pp. 374–392

Chapter 4
Putting practice into policy: reconfiguring questions of consumption and climate change
Elizabeth Shove
Contemporary Social Science, volume 9, issue 4 (December 2014) pp. 415–429

Chapter 5
Input–output analyses of the pollution content of intra- and inter-national trade flows
Karen Turner, Cathy Xin Cui, Soo Jung Ha and Geoffrey Hewings
Contemporary Social Science, volume 9, issue 4 (December 2014) pp. 430–455

Chapter 6
Decentralising energy: comparing the drivers and influencers of projects led by public, private, community and third sector actors
Bouke Wiersma and Patrick Devine-Wright
Contemporary Social Science, volume 9, issue 4 (December 2014) pp. 456–470

Chapter 7
Urban experiments and climate change: securing zero carbon development in Bangalore
Harriet Bulkeley and Vanesa Castán Broto
Contemporary Social Science, volume 9, issue 4 (December 2014) pp. 393–414

For any permission-related enquiries please visit:
http://www.tandfonline.com/page/help/permissions

Notes on Contributors

Wokje Abrahamse works as a post-doctoral fellow at the University of Victoria, Canada. Her research focuses on the effectiveness of interventions to encourage environmentally friendly behaviours, such as energy conservation, travel mode choice, and sustainable food consumption.

Harriet Bulkeley is Professor of Geography, Energy, and Environment at the Department of Geography, and a Deputy Director of the Durham Energy Institute, Durham University UK. She is author of *Climate Change and the City* (Routledge, 2012), and co-author of *Cities and Climate Change* (with Michele Betsill, Routledge, 2003), *Governing Climate Change* (with Peter Newell, Routledge, 2010), and *Cities and Low Carbon Transitions* (with Vanesa Castán Broto, Mike Hodson and Simon Marvin, Routledge, 2011). From 2008 to 2012, she held an ESRC Climate Change Leadership Fellowship, Urban Transitions: climate change, global cities and the transformation of socio-technical systems, and is a coinvestigator on several different projects examining the relationship between energy use and the dynamics of socio-technical systems.

David Canter is an applied social psychologist who started his career in the 1960s as an environmental psychologist at Strathclyde University, Glasgow, UK. Since then he has published the results of his research on a wide range of topics, including the design of offices, hospitals and schools, behaviour in fires and emergencies, and complementary medicine. His 1977 book *The Psychology of Place* is still widely cited. Over the last quarter of a century he has developed the discipline of Investigative Psychology. He is currently editor of *Contemporary Social Science,* the journal of the Academy of Social Science.

Vanesa Castán Broto is a Lecturer in the Development and Planning Unit of the Barlett Faculty of the Built Environment, University College London, UK. Her research looks at urban socio-environmental transformations and the production of environmental knowledge, examining how these processes are mediated by environmental governance mechanisms. She co-edited *Cities and Low Carbon Transitions* (with Harriet Bulkeley, Mike Hodson and Simon Marvin, Routledge, 2011). She is currently leading an action-research project developing partnership agreements for climate compatible development in Maputo, Mozambique (funded by DfiD through the Climate Development Knowledge Network). She is also a co-investigator in other projects investigating the relationship between urbanization, climate change, infrastructure and planning.

Cathy Xin Cui gained her Ph.D. from the University of Stirling, UK. She joined the ESRC Climate Change Leadership Fellowship as a Research Fellow at the University of Strathclyde, Glasgow, UK, in 2010. She continues to work at the Fraser of Allander Institute at the University of Strathclyde.

NOTES ON CONTRIBUTORS

Patrick Devine-Wright is an expert on 'NIMBYism' – the social acceptance of low-carbon energy technologies and associated infrastructures. In particular, he has studied the relevance of people-place relations, specifically place attachments and place identities, to understanding land-use conflicts, informed by theory in Human Geography and Environmental Psychology.

Birgitta Gatersleben is Senior Lecturer in Environmental Psychology at the University of Surrey, Guildford, UK. Her research focuses on understanding the instrumental, affective and symbolic aspects of (un)sustainable consumer behaviour and perceptions of the natural world.

Soo Jung Ha gained her Ph.D. in 2008 after studying at the Regional Economics Applications Laboratory at the University of Illinois, Urbana-Champaign, IL, USA. in 2008, She is now based at the Korean Research Institute for Human Settlements, Seoul, South Korea.

Geoffrey Hewings is the Director of the Regional Economics Applications Laboratory (REAL) at the University of Illinois, Urbana-Champaign, IL, USA.

Catherine Leyshon is a cultural geographer interested in landscape, identity and sense of place. Recently, she has explored these themes in the context of climate change and landscape management in papers in *Progress in Human Geography*, *Climatic Change*, *Area*, *Landscape Research,* and *Geography Compass*. She was part of the UK's National Ecosystem Assessment Follow on Phase (2014), involved in Work Package 4 on Cultural Ecosystem Services. To this, she has contributed a case study on partnership working on the Lizard Peninsula in Cornwall, UK.

Niamh Murtagh is a Research Fellow at the University of Surrey, Guildford, UK. Her work looks at the relationship between self-identity and behaviour, particularly the involvement of self in behaviour change.

Elizabeth Shove is Professor of Sociology at Lancaster University, UK. She has recently completed an ESRC Climate Change Leadership Fellowship on Transitions in Practice: Climate Change and Everyday Life, and is a member of the ESRC Sustainable Practices Research Group. She is the co-author of *The Dynamics of Social Practice* (with Mika Pantzar and Matt Watson, 2012).

Karen Turner is a Reader in Economics at the University of Stirling, UK. She was one of six Climate Change Leadership Fellows funded by the UK ESRC between 2008 and 2010.

Bouke Wiersma is based in the Department of Geography at the University of Exeter, UK, and researches social aspects of energy and sustainability, with a particular focus on public energy deliberations on the local level. His current work explores these in the context of offshore renewable energy in Guernsey.

The challenge of climate change

David Canter

International Research Centre for Investigative Psychology, The University of Huddersfield, Huddersfield, UK

It is generally accepted that human behaviour needs to change if the depredations of climate change are to be reduced. Yet despite overwhelming evidence for this need there has been remarkably little modification of what people do to limit climate change. This book brings together a number of social science studies that demonstrate the challenges and potential for such adjustments to human actions. The studies brought together here support the view that there is a need to look beyond scientific facts about the environment if human environmental activity is to be modified. People modify what they do when there are incontrovertible experiences that demonstrate that the social rules that shape interactions with their surroundings ('rules of place') must be changed. The range of social science studies in this book indicates the great difficulty of modifying human quotidian behaviour without altering the social and societal context that supports it.

Why do we leave it so late?

Most people know the climate is changing in ways that are deleterious. There is even a term of insult that parallels the outlawing of holocaust deniers – climate change deniers. Furthermore, most scientific opinion supports the view that human activity is a dominant cause of these unwanted environmental changes. So why are we doing so little to change our actions in order at least to reduce their impact? As Sunstein (2007) points out the US has responded much more effectively to the threat of terrorism than to climate change. The reason Sunstein gives, as did Cronon (1996) much earlier, is that human cognitive processes are limited in appreciating the real significance of climate change and that, in essence, more information is needed to convince people of the need to modify their activities.

As important as this perspective is it undervalues the social processes that hold back human actions in the face of change. Studies of behaviour in shorter term emergencies and disasters (Canter, 1990) indicate that an understanding of how people make sense of and use their environment indicates that there are crucial limits on what people are prepared to change in regard to their use of their environment.

These studies of behaviour in life-threatening situations demonstrate how wedded people are to their daily activities. The actions in place are not some superficial aspect of a person but are integral to individuals' self-concept. The person we think we are is shaped by the patterns of place use we participate in. As Raymond et al. (2010) demonstrated the attachments people have to where they live incorporate a mixture of views of themselves and their relationships to others.

The processes that introduce inertia into reactions to environmental change, and limit variations in behaviour so that it is not modified to reduce environmental threats, are fundamentally social processes. These are the same processes that have led to many emergencies in the past getting out of control to become disasters, despite early warnings of imminent danger.

The early stages of many disasters are often ambiguous. It is unclear what the threat is or even if there is one. It is this ambiguity that delays serious responses. Why risk seeming to be silly when there is no clear indication that there is anything untoward happening.

The first law of human action

Actions in emergencies illustrate what might be thought of as a law of human activity that parallels Newton's first law of motion. People carry on with their existing behaviour unless some external force leads them to recognise that some new set of rules are now guiding their activities. This gives rise to regular mis-estimates in the development of danger. This is illustrated in Figure 1. There the difference between how people think danger grows and how it typically does grow is represented.

Broadly speaking, it is often assumed that danger grows in a simple, linear fashion even though it tends to develop in an exponential or geometric way. In the early stages, therefore, the estimates of growth are reasonably accurate. This lulls people into thinking that they understand what is happening and it is manageable. But their estimates become ever more inaccurate as time moves on until the emergency gets out of control.

The reasons for this inertia are the existing patterns of behaviour and the understanding of what is expected of interactions with each other and with the known context. It takes a major jolt to re-evaluate what is happening and to change. It is necessary to recognise that the context has changed so that different social rules apply – until that is the case, people are reluctant to change their behaviour.

This behavioural inertia is what happens in small scale disasters where people die in their own homes, major industrial accidents such as Piper Alpha and the *Herald of Free Enterprise*, and even in large scale international disasters such as Rwanda and Bosnia, or the recent Ebola outbreak. Typically early indicators of impending disaster are ignored with the consequent delay in recognising that something needs to be done, until the indicators of danger are so over-whelming that decision makers can then accept that the situation has changed to something so

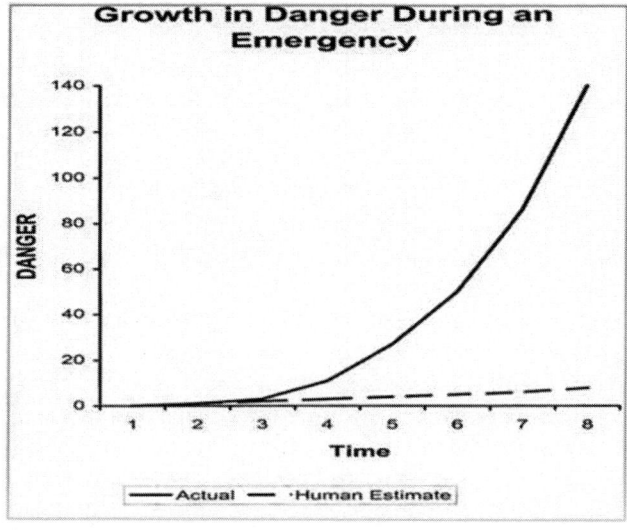

Figure 1.   A comparison of estimates of the growth in danger with its actual growth, showing that under-estimates increase rapidly over time.

different that they can now apply new rules. By then it is often too late and disaster is extremely difficult to avoid.

It is worth emphasising this central point. Emergencies become disasters because initial warnings are ignored. They are not ignored out of obstinacy or ignorance but because of the psychological processes that underlie habitual patterns of activity. The examples are all too familiar. Disasters on the railways came about in a context in which trains were going past signals set at danger. Before the New Orleans floods many people said the levees needed to be developed and strengthened and that more money needed to be put into flood control. The emergency is predictable but the processes in place limit effective responses.

With climate change the ambiguity that characterises the starts of so many disasters is made worse by the fact that it is a concept rather than a direct experience. In her interesting essay in Chapter 2, Catherine Leyshon points out that although climate change is hardly ever out of the news, it is an invention of empirical analysis. It is possible to watch the sun rise, so noting the change in the light, but longer term and subtler changes, whether it be inflation, the average age of the population or global warming, are only noticed by observers who record and measure over what might be long periods of time. Because of this the changes are actually creations of those who make the recordings. They are therefore open to challenge; challenge not just to their existence but, more importantly, what is causing them. In this crucial sense, then, Climate Change is an aspect of knowledge; knowledge that has accumulated over the last 30 years or so. Climate change is not an immediately observable aspect of experience.

Social processes and climate impact

The findings from the study of human reactions in emergencies combined with the abstract nature of climate change point to the importance of understanding the social and psychological issues behind personal decisions and actions. Rather than dealing with people as rational automata to be manipulated by economic and political forces, they need to be regarded as sentient agents interacting consciously with their environment.

There may be global changes that create local conditions, but it is their impact at the level of individuals going about their daily activities that produces their effects. If a micro-climate manages to avoid the global pattern then actions and organisms in that location will not be effected. It is necessary to consider what it is that maintains activity in any given setting in order to understand why people leave it so late to deal with environmental threats.

A further understanding of how local processes, relating to interactions between individuals, is the key to environmental change can be gleaned from a consideration of the roots of present day concerns with the use and abuse of the environment. It is not usually appreciated, as Cronon (1996) indicated some time ago, that the environmental movement and concerns with global changes can be traced to roots in the early Romantic Movement and the Romantic poets. This predates Darwin's *Origin of Species* by almost 100 years.

It is an intriguing aside to note that it is often poets, novelists and playwrights who identify crucial issues and draw attention to them before scientists eventually turn them into something more mundane and technical, often losing some of the emotional power of the poet's insight. This may be one of the messages for those who seek to have climate change taken more seriously, perhaps what is needed in the debate is more drama and poetry that deals with localised, individual experiences and fewer facts and figures?

Looking back to what is often regarded as the origins of concerns with the environment there are some remarkable parallels to present day discussions. Oliver Goldsmith's (1730–1774) poem, *The Deserted Village*, published in 1770 is regarded by many as the starting point for the development of a Romantic attachment to the natural, rural landscape and a notional Golden Age in which humanity all lived in a pleasant, Arcadian environment. The essence of the poem is captured in the stanza:

> Sweet, smiling village! Loveliest of the lawn
> Thy sports are fled, and all thy charms withdrawn:
> Amidst thy bowers the tyrant's hand is seen,
> And desolation saddens all thy green;

He is bemoaning the fact that this idyllic village, Auburn, in which he grew up is now empty. The people have left the countryside to go into the nasty town. The trend, started most clearly with the British industrial revolution, is today mirrored across the globe, with peasant, village life fading as cities develop rapidly with vast numbers of individuals coming in from the countryside. And who is causing the problem; the tyrants, the rich multi-nationals; the organisations that are making it more attractive to go into the town.

Although *The Deserted Village* is apparently about the loss of a rural idyll and the degradation caused by urbanisation it is more fundamentally an exploration of the breakdown of the relationship between people and their physical environment. It is an early recognition that countryside is not just a resource to be exploited but an integrated part of who we are and how we live.

In many senses the poem heralds a new humility in which the domination of nature by humanity is no longer the unchallenged norm. It is a budding flower of the Romanticism, given such impetus by Wordsworth and Coleridge, in which people are seen as part of nature rather than its inevitable masters. The Romantic vision that perfection resides in the everyday, rather than in some aspired to ideal, puts human beings on a par with nature rather than being above it. The ecological implications, intriguingly, that we owe our survival as a species to how we interact with our environment, has some of its origins in concerns for the consequences of leaving villages for the depredations of the town.

The complexity of environmental concerns

As noted, climate change is not directly available to the senses – it is an idea, a construct. Like many ideas that emerge into public consciousness it covers many different things; everything from changes in the icecaps, to reduction in biodiversity, depletion of natural resources such as water or land or energy. Debates on environmental degradation also become embroiled in explorations of policy options, challenges to the economies of developing countries and the gamut from education to forms of burial. With such a diversity of issues under one umbrella term it is perhaps not surprising that it is so difficult to generate actions to deal with them. What exactly is the phenomenon that needs to be addressed?

It was Winston Churchill who summarised the complexity of the interaction with the environment with his characteristic eloquence. During the Second World War, after the Houses of Parliament had been bombed, he led the debate on the re-building saying:

> On the night of May 10, 1941, with one of the last bombs of the last serious raid, our House of Commons was destroyed by the violence of the enemy, and we have now to consider whether we should build it up again, and how, and when. We shape our buildings, and afterwards our buildings shape us. Having dwelt and served for more than forty years in the late Chamber, and having derived very great pleasure and advantage therefrom, I, naturally, should like to see it restored in all essentials to its old form, convenience and dignity.
> 28 October 1943 to the House of Commons (meeting in the House of Lords).

His central argument, that was accepted when parliament was rebuilt in 1950, was that the old House of Commons should be recreated as it had been, even though there were not enough seats for every Member of Parliament to have a place, as well as a number of other inefficiencies. Churchill argued against any increased efficiency saying that democracy has been threatened in other countries by 'giving each member a desk to sit at and a lid to bang'. As he explained, on many occasions the Chamber of the Commons would not be very full, so the smallness would maintain intimacy, whereas, at critical times it would be overcrowded, with members pouring into the aisles, supporting a 'sense of crowd and urgency'.

Churchill thus articulated two crucial aspects of the experience of places. One is that often the symbolic qualities, the meanings that are assigned to what is going on, outweigh any simple functional analysis. The second is the ongoing interaction between places and their use: *We shape our buildings, and afterwards our buildings shape us.*

There is no simple one-way process from environment to response, or from response to environment. There is always an unfolding dynamic system; transactions between creating an environment for what we want and how we want to use it gives rise to a sequence that leads us to then interact with that place in the way that we expect. This is a powerful and complex dynamic.

It is the power of this dynamic interaction that explains why so often we leave it so late. We find it very difficult to break out of existing habits that are structured and in their turn structure how we make use of our surroundings. They are an integrated part of who we think we are and what we see as our intentions that give meaning to our actions. They all derive from our transactions with places, and the rules that guide them; 'Rules of Place' (Canter, 2013). Simply insisting that people turn down their thermostats, or don't fill their kettles for one cup of tea, or fly less often is not enough without setting in motion a cycle that changes our whole way of interacting with and within places.

Psychologists, of course, have often been aware that people do not always just act logically on the basis of evidence. In relation to pro-environmental behaviour Birgitta Gatersleben and her colleagues demonstrate in Chapter 3, reviewing a number of large scale studies, that who people think they are – their identity – is tied up with what they do. Such identities come from many sub-cultural and personal-historical factors. Indeed, as Catherine Leyshon argues in Chapter 2, policies attempting to changes human actions are built on assumptions about the nature of people. Are people just economic entities influenced by the balance of costs and benefits? Or are they better thought of as social beings, shaped by their interactions with others.

Elizabeth Shove in Chapter 4 discusses the problem of enabling policy makers to engage with this social science perspectives. She shows that by recognising the cultural, place specific, context that defines acceptable behaviour quite dramatic impact is possible. She quotes the Japanese example of encouraging salarymen to take off their jackets and ties in summer as a way of reducing the need for air-conditioning. In the tight codes of the Japanese office workspace it required this formal approval of a different dress code to generate a remarkable reduction in CO_2 emissions.

That Japanese 'Cool Biz' policy illustrates the complex processes involved in dealing with climate change. Social habits, technological systems, and policy formulations are all of relevance to human biological processes, all had to be harnessed together to make a difference. It is the growing awareness of the need to understand and manage multi-modal, multidisciplinary systems that social scientists are articulating as they develop their explorations of climate change and what to do about it.

The awareness that threats to our environment are part of complex social and economic systems can nonetheless generate some simple implications. In a sophisticated economic analysis of the pollution content of the flows of trade Karen Turner, and her colleagues, in Chapter 5, demonstrate that managing the consumption can be as, or even more effective, than managing production. This has obvious parallels to dealing with drug abuse, in which reducing drug use may be much more powerful than trying to stop production.

Behaviour change starts at home

The recognition that energy uses are part of complex systems of supply and demand, within a socio-cultural context, has been one of the motivators for encouraging smaller and tighter systems. Some years ago the distinguished anthropologist Robert Rapport (2001) argued that our relationship to the environment will only change if the appropriate values systems are developed within the family. Decentralising energy production so that it is closer to where it is consumed accords with this advocacy. By having the energy generation as part of a local community makes it much more viable to incorporate family values when considering production and consumption.

The development of local energy production is seen as a way of not only reducing CO_2 emissions, but also of increasing energy security and engaging the grassroots in the whole production and consumption cycle. Bouke Wiersma and his colleague, in Chapter 6, review a number of studies of decentralised energy projects. They included community installed bio-mass fuelled district heating system, the construction of zero-carbon homes and local wind-turbines. What became clear was that energy efficiency was typically only a small feature of the reasons for these projects. Fuel poverty, improving housing quality standards, health and well-being were typically more dominant in influencing the reasons for the projects.

Such findings serve to show how the examination of the social processes involved in dealing with climate change reveal that those processes usually encompass much more local issues than the grand ideal of saving the planet.

The interrelationship of the context of change with idealistic missions is also well-illustrated in the novel study carried out by Harriet Bulkeley and her colleague, in Chapter 7, of a zero carbon development in Bangalore. They show that making, maintaining and living low carbon are three inter-related activities. As such they have a tendency to isolate the development from its community. So, although the T-Zed project as a carbon-neutral gated community is transforming and reconfiguring the urban landscape of Bangalore, its impact is also deeply ambivalent because of the way it sustains existing forms of urban development and the associated inequality.

The psychology of place

The Bangalore study draws attention to the importance of understanding the geographical context of attempts to deal with climate change. But this understanding benefits from awareness that places are not just physical locations. They incorporate many social and psychological meanings.

In English this is illustrated by the word 'place' having a rich and abstract set of meanings that are not easily translated into other languages. It can imply a position in a conceptual hierarchy, 'knowing your place', a very small and specific location, such as your place at table, or a much larger town square, or even a city 'the place of your birth'. It certainly carries meanings beyond a mere physical location that is often lost with translation into other languages. You cannot use the Japanese word for place, 場所, in the sense of reducing overweening confidence, as in the phrase 'putting a person in his place'. Although interestingly the Hebrew translation מקום can also take on the meaning of God in the sense of 'The Omnipresent', reflecting the symbolic qualities inherent in the idea of a location that is more than just a point in physical space.

In an attempt to grapple with the subtleties of the meaning of 'place' a more technical definition was developed (Canter, 1977). This proposes that places are composed of three interacting components;

- the activities that give meaning to a location,
- the physical form of that setting, and
- the conceptualisations associated with those activities in that context.

This sense of place is what Goldsmith was reaching for. It is this concept of particular types of natural places that define the community which have a social psychological, a physical and an active component to them. It is these conceptual, physical and use components which gives places meaning and create the pattern of use that are so difficult to break. How places are used is not casual or arbitrary. It is part of a complex, evolving set of social norms and personal habits that are very difficult to modify, being a major cause of the lethargic response to climate change.

References

Canter, D. (1977) *The Psychology of Place*. London: Architectural Press

Canter, D. (1990) Studying the Experience of Fires, in D. Canter (ed) *Fires and Human Bhevaiour: Second Edition* London: David Fulton Publishers 1–14

Canter, D. (2013) Why do we Leave it so Late? Response to Environmental Threat and the Rules of Place. *Journal of Earth Sciences and Climate Change* 5: 169. doi:10.4172/2157-7617.1000169

Cronon, W. (1996) The Trouble with Wilderness: Or, Getting Back to the Wrong Nature *Environmental History 1 (1) 7–28* http://www.jstor.org/stable/398505910.2307/3985059

Rapoport, R. (2001) *Families as Educators for Global Citizenship* Farnham: Ashgate Publishing

Raymond, C.M., Brown, G., Weber, D. (2010) The Measurement of place attachment: Personal, community and environmental connections. *Journal of Environmental Psychology ,*30, 422–434. doi: 10.1016/j.envp.2010.08.002

Sunstein, C.R. (2007) On the Divergent American Reactions to Terrorism and Climate Change *Columbia Law Review* 107, (2) 503–557

Critical issues in social science climate change research

Catherine Leyshon

Department of Geography, University of Exeter, Penryn, Cornwall, UK

This paper examines the challenges and opportunities for social scientists working on climate change research. Much work is required to expose and destabilise taken-for-granted assumptions about: (i) the nature of climate change, its complex ontology and knowledge-making practices; and (ii) how academic knowledge is made at the expense of other ways of knowing, doing and being in the world. I examine the relationship between the natural and social sciences, the epistemological question of what people are, and the multiple spaces, sites and practices across which and about which social science research on climate change is being produced.

Introduction

In this paper I will make a series of linked points about the challenges and opportunities for social scientists working either directly or indirectly on climate change research. Although much greater strategic effort – by the International Social Science Council (ISSC) in particular (but see also Castree et al., 2014) – is going into positioning the social sciences[1] as an indispensable part of reframing and understanding climate change as a social phenomena, there remain many questions about how to mobilise a large and disparate field in a common cause. In this paper, I examine some ontological and epistemological issues as well as recent research on spaces, sites and practices. I argue that we must go further in examining the nature of the social science enterprise itself if we are to use its many intellectual resources to solve the world's most pressing problem. To begin, I examine critically the issue of what we are talking about when we talk about climate change by exploring its ontological politics.

Ontological politics of climate change

Ontology is the study of being concerned with questions about what actually exists in the world that humans can acquire knowledge about. Writing in the context of conservation biology (though the point holds for climate change research), Moon and Blackman (2014) argue that ontology is important because it helps researchers recognise how certain they can be about the nature or existence of objects they are researching. Such choices have consequences for theory, practice and normative positioning (Goodman, 2001) but they are also political, defining intellectual priorities, approaches and solutions. There are two issues in the ontological politics of climate change that I wish to raise here. The first is the problem inherent in climate change itself and the second – blurring into epistemological issues – is the consequences of thinking about climate change in terms of a coupled human–natural system or socio-ecological system. Both the positions I discuss below have implications for almost every aspect of what we study under the banner of climate change research, how we study it and how we communicate it to disparate audiences, and yet they are rarely subjected to any sustained scrutiny in the academic literature (Esbjörn-Hargens, 2010).

Climate change

At the risk of generalisation, the natural sciences are dominated by realism – the idea that a reality exists that can be studied, understood and experienced as truth. But realist climate science has a contaminant at its heart: climate itself. For climate change is not an object. It is a category of knowing (Brace & Geoghegan, 2011) made out of approximately 30 years' worth of meteorological observations including surface variables such as temperature, precipitation and wind (Hulme, Dessai, Lorenzoni, & Nelson, 2009). Naustdalslid (2011, p. 243, original emphasis) observes that

> climate change is 'man-made' ... in the sense that it is only visible to man [sic] and society through science ... without climate research, without the concerted action of scientists under the IPCC and without the systematic and convincing dissemination from this scientific activity to policymakers and the public, *climate change would not have been visible as a problem for society today.*

Thus, Leyshon and Geoghegan (2012, p. 237) identify what they call a 'metaphysical and semiotic problem' with climate change, exposing the ontological tangle that occurs when climate as a statistical construct assembled from a range of meteorological, oceanographic and atmospheric data is 'treated like a homogenous entity on a trajectory of change towards an altered state on a timescale that exceeds the life expectancy of today's primary school children'. Esbjörn-Hargens (2010, p. 144) characterises climate change as a hybrid ontological object – 'a combination of scientific third-person observations and cultural second-person meanings'. Climate change is, in Brace and Geoghegan's terms (2011), made of the stuff of everyday life (such as weather), but is not in and of itself that stuff. Thus, although we might experience some warmer summer days, suffer longer droughts or observe glaciers retreating, climate has few consistently affective qualities that confirm unequivocally to those outside the science community that change is happening.

The ontological status awarded to climate change in the natural sciences is quickly destabilised by relativism, and it is these same relativist perspectives that help us to see why, despite overwhelming scientific evidence, climate change has not generated, *inter alia*, widespread public action, behaviour change, coherent local, regional, national or international policy, or successful regulatory tools. Climate change is simultaneously a reality, an agenda, a problem and a context but one with uncertain imminence (Brace & Geoghegan, 2011; Leyshon (née Brace) & Geoghegan, 2012).

The sheer range of topics and environmental issues that are studied under the banner of climate change both reflects and compounds the metaphysical and semiotic problem (Box 1). These also demonstrate that climate change is not mutually exclusive from other pressing global environmental issues. As the World Social Science Report 2013 points out, climate change is only one of a suite of linked global environmental challenges that encompass all the biophysical changes happening on land, in the oceans, and in the cryosphere and atmosphere

(ISSC/UNESCO, 2013) and which include biodiversity/habitat loss, energy security/peak oil, population growth, ocean system collapse and water degradation. The nature of these conjoined biophysical and social problems required a new conceptual framing, to which I now turn.

Box 1. Prominent topics and themes in social science research (Hackmann and St Clair 2012, p.15).

- Central issues of climate change impacts, adaptation, mitigation, vulnerability, resilience and sustainability.
- Concerns related to ecosystems, environmental services and biodiversity.
- Problems of primary resource depletion and needs related to water, energy, land, food and so on.
- Population growth, migration, displacement to urbanisation, waste management, oceans and coastal vulnerability, extreme events, disaster risks, social protection, peace, security and conflict, poverty, inequality, governance, innovation technological assessment.
- Sector-specific priorities: development pathways, green growth, education, media, health, agriculture, law, international relations, transport, science policy.
- Policy and response: clean development mechanisms, geo-engineering, economic initiatives, developing country-focused programmes.

Coupled human–natural systems

It has become increasingly popular to talk about 'coupled human–natural systems' – sometimes also phrased as a 'socio-ecological systems' or 'human–environment systems' – when discussing global environmental issues. The dominance of the 'system' as a way of conceptualising the world has its modern roots in the Enlightenment, through, for example, Thiry's *The System of Nature* (1770). One consequence of this way of thinking has been to focus on the component parts of the system, often through disciplinary specialisms which have, over time, neglected an exploration of the complexity of the whole earth system in favour of every more forensic and detailed examinations of its component parts (Funtowicz & Ravetz, 1993; Lowe, Phillipson, & Wilkinson, 2013). In climate science, concepts like 'feedback' and 'tipping points' suggest a recognition that some aspects of the interactions of the 'system' are not sufficiently well understood to be able to accurately predict their outcomes (Hall & Pidgeon, 2010).

There is clearly no appetite to abandon the dominant ontology of the system in the face of the unprecedented challenge of climate change, but there are calls to work differently to understand the system. There are repeated assertions that, as systems combine 'human, biological and physical elements that link together diverse people, places and processes through multiple material flows and intermediaries' (Lowe et al., 2013, p. 213), they must be studied in an interconnected way. There is growing recognition that, rather than study the individual parts in isolation, the *relationships between* the individual parts may be more important, as are the processes that effectuate the system as much as its structures. Further, Popa, Guillermin, and Dedeurwaerdere (2014, p. 2) show that, in contexts where there are 'a plurality of decision-makers, pervasive uncertainties, spatial and intertemporal externalities, interplay of human and natural components and an evolving understanding of policy objectives' the problem becomes one of managing complex social-ecological systems under conditions of uncertainty and with a plurality of values and perspectives.

However, as academics from across the natural and social sciences strive to haul the whole system into view, in all its magnificent complexity, they are driven by a familiar Enlightenment belief in the ability of reason to illuminate not only problems but solutions to our world's most pressing crisis. As the recent ISSC/UNESCO World Social Science Report (2013, p. 7) on changing global environments asserts, 'approaching global environmental change from a systems perspective draws attention to nonlinear relationships and the potential for irreversible changes and surprises'.

Thus, the reliance on the system represents what we might call academic path dependence, particularly in the way that a broad research community seems to be locked in. One consequence of this, as Castree et al. (2014, p. 764) note, is that a particular framing of the human dimensions of climate change has become normalised.

The frame's major presumption is that people and the biophysical world can best be analysed and modified using similar concepts and protocols ... A single, seamless concept of integrated knowledge is thereby posited as both possible and desirable, one focused on complex 'systems'. The frame positions researchers as metaphorical engineers whose job it is to help people cope with, or diminish, the Earth system perturbations unintentionally caused by their collective actions.

Having said this, there is evidence in the idea of a coupled human–natural system of a modest ontological evolution away from what might be described as Enlightenment ways of thinking in which the social or cultural realm of human endeavour was symbolically separated from the natural world. The conjoining of the human and the natural in the phrase 'human–natural system' works in several ways as a heuristic. First, it goes a little way to dissolving a metaphysical boundary between artificially constructed realms (something that geographers have been attempting by working with human–non-human relations and more-than-human geographies, see, for example, Whatmore, 2006). At worst, however, the idea of a coupled human–natural system is used uncritically to simply describe a system in which people interact with natural components (Liu et al., 2007).

Second, the phrase signals a recognition that humans have responsibility for widespread and potentially irreversible alterations to the planet's biophysical processes (which has also engendered a new framing – the Anthropocene – and a new journal *The Anthropocene Review*, Oldfield et al., 2014). Third, it implies something about possible solutions to the problems that are of our own making and need a much greater understanding of the system and our place in it from across ontological, epistemological and disciplinary communities to improve policy, management and governance of the environment (Liu et al., 2007). As Castree et al. (2014) note, this is much easier to aver than achieve. Finally, the phrase 'human–natural system' and its variants hint at the complexity of the issues that face us – the so-called 'wicked problem' of climate change. For Naustdalslid (2011), climate change represents a modern environmental problem which 'normal science' cannot resolve. He is not alone in his appeal to post-normal science as a means of increasing the salience, credibility and legitimacy of science to make it more useful in practice (Knapp & Tainor, 2013).

Like other recent heuristic devices – such as Ecosystem Services (Leyshon, 2014) – the idea of a coupled human–natural system is not one which achieves immediate resonance beyond academe. Further, despite some now quite strident assertions of the value of the social sciences in understanding and responding to climate change through the heuristic of the coupled human–natural system (Hackmann & St Clair, 2012; ISSC/UNESCO, 2013), it is a rhetoric that is perhaps proving difficult to realise in practice.

Epistemology

In this section I consider the nature of the dialogue between the social and natural sciences, and the fundamental epistemological issues that must be faced if the social sciences are to fulfil their potential in tackling climate change.

The social and natural sciences

Lowe et al. (2013, p. 207) have recently observed that it 'has become part of the mantra of con-temporary science policy that the resolution of besetting problems calls for the active engagement of a wide range of sciences' including the social sciences. This, Lowe et al. (2013) argue, has become necessary because the nature and ubiquity of environmental change has called attention to the contingency of the natural (as well as the social) and undermined belief in the permanence of the natural world. However, the enrolment of the social sciences has been nowhere near straightforward for a couple of reasons. First, the problem is urgent: the climate science is unequi-vocal and changes to individual and societal attitudes and behaviours must be enacted *now*, in accordance with international agreements and previous Intergovernmental Panel on Climate Change (IPCC) reports. Prevarication seems like an indulgence. The rhetoric of urgency militates against the sort of reflexive consideration of social science's means and ends. Second, and in a linked point, the historical dominance of the natural sciences in framing climate change research conditions the terms of entry for other disciplines.

Moon and Blackman (2014) propose a guide to social sciences research for natural sciences that will achieve three aims: (i) understand the philosophical basis of the social sciences; (ii) inter-pret social science; (iii) appreciate alternative approaches to scientific inquiry. Their purpose is to 'open the door' to social science research and assert the legitimacy of its principles, assumptions and interpretations (Moon & Blackman, 2014, p. 2). The very fact that these authors feel the need to assert the legitimacy of a field of enquiry that has been in existence for at least the last two hundred years illustrates and also does nothing to challenge the dominance of science at the top of a hierarchy of ways of knowing.

It is not only the absence of knowledge about the intricacies of social science that prevents natural scientists engaging. It is also, as Lowe et al. (2013, p. 208) suggest, the 'casting of social science in an end-of-the-pipe role', as the translator of scientific and technological devel-opments (see also Hackmann & St Clair, 2012). Thus, Lowe et al. (2013, p. 207) suggest that 'social scientists have typically been forced into an auxiliary role of supporting and interpreting developments in natural science and technology'. Such an auxiliary position is still evident in Weaver et al's assertion (2014, p. 256 – emphasis added) that a key role of the social sciences is to '*elucidate* the processes that turn knowledge into action' and enable collaborations and dia-logues between scientists of all kinds and practitioners, producing 'effective, science-based decision-support for global change-related problems' and 'knowledge that is practically relevant, usable, credible, legitimate and actionable'. Castree et al. (2014) argue that the potential fruits of interdisciplinary exchange are not only far greater than those being propounded in various reports but different in character. They suggest that researchers interested in global environmental change should consider different 'values-means-ends' packages, wherein:

> values are those fundamental beliefs that motivate people's behaviour (for example, love of nature, the right to free speech); means are those various practices, procedures, institutions and technologies by which values can get instituted; and ends are the concrete goals to which means are orientated and which provide a measure of how well values are being realized at any one time or place. (Castree et al., 2014, p. 766)

This would position researchers across disciplines as working together to 'open up the range of choice available to societies' and 'rather than assuming that one form of broad-based, integrated, actionable knowledge 'fits' any given situation, researchers would together make visible a number of actual and possible realities'. This may be very discombobulating for some physical scientists who have 'grown accustomed to a certain "style" of human dimensions research' (Castree et al., 2014, p. 766) which they find easier to accommodate than approaches which are broadly post-positive, critical and interpretative. Leyshon (2014) identifies this as 'epsitemic distance decay' wherein the social science disciplines with at least some recognisably scientific ontological, epistemic and methodological concerns feature most strongly as collaborators with the natural sciences.

Leaving aside the question of how the dialogue between the natural and social sciences should be taken forward, there is also the issue of how the social sciences represent their endeavour. For Hackmann and St Clair (2013), the way forward is in joint, reciprocal framing, mutual learning and then the co-design, execution and application of research. To facilitate this, they set out six 'transformative cornerstones' of social science research that describe the unique capabilities of the social sciences. They are transformative because they 'work together to inform action for deliberate transformation that is both ethical and sustainable' (p.16).

Box 2. Transformative cornerstones of social science research (Hackmann and St Clair 2012, pp.16–20).

Cornerstone 1 – Historical and contextual complexities – distinguishing multiple stressors, drivers and interdependencies; learning from history; dealing with differences across geographical, cultural, personal, professional contexts and identities.

Cornerstone 2 – Consequences – living with global change: taking stock of threats and impacts across different groups and regions; identifying social boundaries and tipping points; measuring success: improving the outcomes of specific actions and instruments.

Cornerstone 3 – Conditions and visions for change – understanding how we can change behaviour and social practice; speeding and scaling up processes of change; building consensus on the directions for change.

Cornerstone 4 – Interpretation and subjective sense making – understanding the nature and role of subjectivities; exposing blindspots; explaining scepticism, indifference and denialism.

Cornerstone 5 – Responsibilities – foregrounding normative agendas; fostering global and intergenerational solidarity and justice; safeguarding ethical approaches.

Cornerstone 6 – Governance and decision making – coming to grips with policy processes and political will; making knowledge work; building relevant institutions and structures.

Impressive though the scale of ambition set out by these cornerstones, such agendas repeat for scholarly audiences the very mistake made in trying to shift public attitudes and behaviours in the face of climate change: the assumption that knowledge straightforwardly and unproblematically produces change. This is a model – grounded in falsification and replicability – that has successfully led to advances in scientific thought and practice with benefits to society. However, the problem of climate change seems stubbornly resistant to this approach.

The transformational cornerstones present what their authors consider to be an irresistible case to a natural science community, predicated on the idea that the culture of science and research will be shifted by the clarity and precision with which the case is presented. The cornerstones are designed to alter 'the fundamental attributes of the system, including ... structures and institutions, infrastructures, regulatory systems, financial regimes, as well as attitudes and practices, lifestyles, policies and power relations' (Hackmann & St Clair, 2012, p. 16). The task set by the social sciences is no less than 'to innovate in ways that lead to new social relations, new social understandings of and responses to the challenge of global change, and new revolutions in socio-economic, political, scientific, educational and legal systems and institutions' (p. 16). But, as Shove (2012, p. 416) cautions in this special issue, 'social theories do not lead directly to prescriptions for action'.

There is little here to challenge the Enlightenment inheritance of the power of reason and rationality. Indeed, the enormity of the task is only matched by the belief in the ability of western-knowledge-making to deliver on it: to make the complexity of the world visible and knowable, to aggregate, scale-up, typify, create typologies, categorise and to therefore engender change on a grand scale, as envisaged by Weaver et al., (2014, p. 657):

> Moving beyond individual knowledge producers and users, we need to understand decision-support processes in the aggregate. We should identify typologies of users, develop comparative studies of decision processes across contexts and scales, and rigorously evaluate the success of such processes.

Even as the World Social Science Report suggests that social transformation is not well understood, the authors are asserting the importance of 'people's capacity to imagine futures that are not based on hidden, unexamined and sometimes flawed assumptions about present and past systems' (ISSC/UNESCO, 2013, p. 9). The social sciences seem to be in the process of creating a narrative in which they seek to be accommodating the dominant ontologies of science (human–natural system) and in which they seem to feel the need to simplify its diversity in order to become accessible to natural sciences. The scale of the climate change problem might be so pressing that we have to smooth and simplify in order to be accepted. We promise to offer insights into complexity without chaos. Castree et al. (2014, p. 766) caution against pulling any punches: 'Framing the 'offer' in terms that meet the … expectations of many physical scientists will inevitably perpetuate the truncated perception [of the environmental social sciences and humanities] we are questioning here'.

Moon and Blackman's (2014) call for natural scientists interested in social science to understand its philosophical principles and theoretical assumptions contains the implicit assumption that social scientists are transparent about such matters. However, as the literature reveals, social scientists rarely make their philosophy and theoretical position clear because: (i) publishing and presentation conventions in general do not demand reflexivity; thus few scholars are required to be absolutely explicit about the ontological and epistemological foundations of their work; (ii) following from this, assumptions go unexamined because they are shared by disciplinary colleagues. Social science is not a single discipline and we should not talk about it as if it was a homogenous intellectual endeavour with shared and stable philosophies, theories or methods. Indeed, despite a growth in the gross number of articles on climate change within the social sciences (Figure 1), the history of scholarly publishing on climate change in social science disciplines from 2000 to 2010 demonstrates that environmental studies, economics and geography have

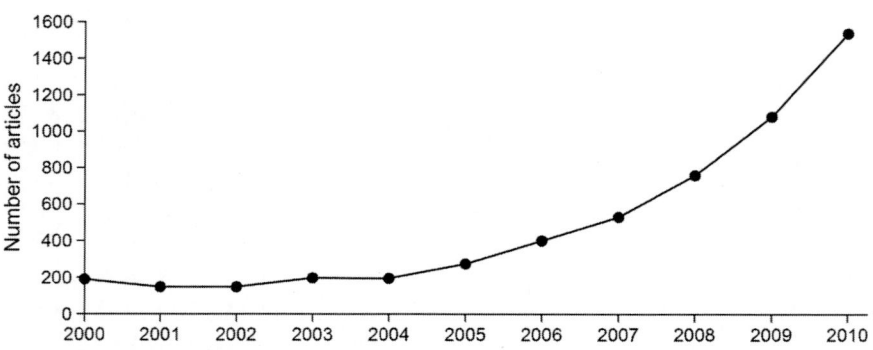

Figure 1. The number of articles with these keywords has grown exponentially in the period 2000–2010 (R^2 =0.905).

dominated whilst scholars from other disciplines with significantly different epistemic and methodological concerns have published fewer papers on the topic (Figure 2).

One of the fundamental epistemological questions that separates different disciplines in the social sciences and which is of importance for climate change research is simply 'what are people like?'. I explore this in the next section.

What are people like?

The answers to the question 'what are people like' range from rational maximisers of self-interest (homo economicus), social subjects of particular discourses (a structuralist perspective) and subjective agents engaged in relational negotiations, improvisations, practices and performances (a view informed by critical theory) – and many variants on these (Gregory, Johnston, Pratt, Watts, & Whatmore, 2009). 'What are people like' is possibly the most critical question for social scientists of climate change because it influences every other question we as social scientists seek to answer: why has not knowledge of climate change altered the way we live in the West? Why do people do what they do, think what they think? Why do they not act in accordance with their beliefs? How can we change ourselves and our societies to adapt and mitigate to climate change? How can we inform successful policy-making for climate change? And, the problem that currently seems to be occupying the minds of leading social science organisations and funders (such as ISCC), how can we work together on climate change research? These and other questions are the foundations of work on climate change as a social-cultural phenomenon (McCarthy, Chen, López-Carr, & Endemanõ Walker, 2014).

In our epistemic communities, our answer to 'what are people like' rarely requires articulation to our peers and this neglect of reflexivity means that working assumptions about values,

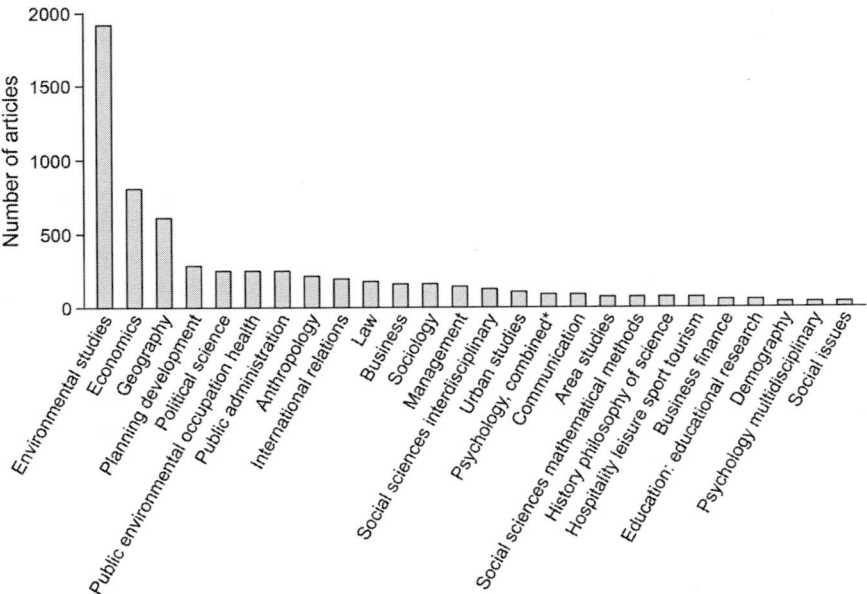

Figure 2.   Only social science (WOS) subfield categories with minimum 30 publications are included. Psychology, combined* is a custom made discipline consisting of a combination of: Psychology experimental or Psychology multidisciplinary or Psychology social or Psychology applied or Psychology clinical or Psychology educational or Psychology biological or Psychiatry or Psychology.

identities, behaviour, society, the individual and the nature of change remain implicit and unchallenged. Yet at least some of the concepts at large in the literature have very different readings in different disciplines. To take but one example: identity. The concept of a person as self-sustaining entity possessed of the capacity of conscious reason and with a core that is essentially fixed and continuous is a product of Enlightenment thinking (Gregory et al., 2009). Whilst it has been modified by research which attempts to retheorise identity as (variously) fluid, fractured, unstable, mobile or as a narrative achievement (Gregory et al., 2009), in some social sciences identity is often still seen as a stable site from which interactions with society (as a distinct entity) proceed. From psychology, Gatersleben et al (2012, p. 377) speak of values and identities as 'generally stable factors that transcend specific situations' whilst from cultural geography Geoghegan and Leyson (2012; see also Geoghegan and Leyshon, 2012) start from the position of identity as relational, practiced, performed and represented. Nagel (2012), meanwhile, examines the significance of gendered identities for understanding the relative vulnerability of men and women to the impacts of climate change.

Such a divergence makes a tremendous difference to the analysis and interpretation that can be brought to bear on empirical data. Gatersleben et al. (2012, p. 374), for example, argue that 'stable individual factors such as values and identities … affect a wider range of behaviours' (see also Jaspal, Nerlich, & Cinnirella, 2014). Leyshon and Geoghegan (2012) on the other hand focus on the embodied, experiential processes and practices around which social meaning is made.

The ISSC/UNESCO World Social Science Report (2013, p. 7) argues that:

> Critical to a social-ecological systems perspective is the role of humans as reflexive and creative agents of deliberative change. Understanding how values, attitudes, worldviews, beliefs and visions of the future influence systems structures and processes is crucial.

If, as a thought experiment, we replace the words the words 'reflexive', 'creative' and 'deliberative' with 'unconscious', 'destructive' and 'thoughtless', we start to see that humans are commonly framed – perhaps even reified – in the social sciences as excogitative, rational agents whose unconscious, habitual, or even wilfully lethiferous behaviours go largely unexamined. Meanwhile values, attitudes, worldviews and beliefs are treated a separate, stable, knowable drivers of behaviour. Dominant approaches from, *inter alia*, economics, psychology and sociology forestall much consideration of the fugacious, commingled qualities of values, attitudes and beliefs or how they inform behaviour. As Shove observes (2012), there is a common understanding of behaviour as something that is driven by rational self-interest, attitude/motivation or habit. There is also an assumption that actors themselves will be able to unproblematically articulate their values, attitudes and beliefs. However, as Kobayashi and Mackenzie (2014, p. 229): suggest,

> action, and particularly the culture-building routines of everyday life, are none the less commonly taken for granted and opaque to actors … the causes of action (or inaction) are not always apparent to actors themselves.

The expectation that people as rational agents will always be able to articulate and account for their actions is to overlook not only people's messiness and complexity but also their capacity for creative and imaginative thinking, doing and making – see, for example, Paschen and Ison (2012) on narrative approaches which 'story' climate change. Yusoff and Gabrys (2011) and Gabrys and Yusoff (2012) call for more attention to climate as a dynamic cultural force capable of reshaping societies and environments. To fully appreciate and engage this approach, it is necessary to take seriously the power of the human imagination 'as a way of seeing, sensing, thinking and dreaming the formation of knowledge, which creates the conditions for material interventions in and political sensibilities of the world' (Yusoff & Gabrys, 2011, p. 516). They identify three distinct temporal and spatial imaginative framings of climate change: the future, everyday life and science–art collaborations. While, as Brace and Geoghegan (2011) note, the future is the *sine qua non* of climate change, crucial to the scientific practices of modelling and prediction, it has also provided the imaginative fuel for catastrophic renderings in art, literature and film of abrupt climate change. Meanwhile, an imaginative recasting of climate change as something that is not 'out there' but 'in here' has engendered a tacit acceptance of the need for adaptive strategies embedded in everyday life. Finally, science–art collaborations have sought to reconsider the 'social spaces of climate interaction and the science–policy–public interface' (p. 517).

Such research innovations clearly draw on the epsitemologies, theories and methods of the humanities to understand human creativity. However, as Castree et al. (2014, p. 765) argue, the environmental humanities have much more to offer, addressing

> fundamental questions of value, responsibility, rights, entitlements, needs, duty, faith, care, government, cruelty, charity and justice in a world marked by (1) significant differences in people's customs and aspirations, (2) manifest inequalities in people's living conditions and material prospects, and (3) complex material and moral interdependencies among people and non-humans stretched across space and unfolding through time.

In the next section, I wish to call attention to the multiple spaces, sites and practices across which and about which social science research on climate change is being produced.

Spaces, sites, practices

Spaces, sites and practices are material and conceptual – for example, research on climate change in urban areas is not just about the urban as a space but the city as a discursive site, and urban governance as a practice. Thinking about spaces, sites and practices helps us to recognise that everyday life is conducted in specific places and through different, sometimes highly routinised, performances which are themselves the product of discursive regimes, constellations of political and personal power, governance structures, regulation and so on. Highly reductionist social science on climate change has tended to elide this complexity but it is now more widely recognised that it must be embraced rather than evaded (Geoghegan & Leyson, 2012; ISSC/UNESCO, 2013).

Although, as noted above, climate change research in the social sciences has tended to see the social, the technological, the economic and the cultural as separate realms, new research is providing more nuanced insights through well-theorised empirical work. Bulkeley and Castán Broto (2012), in this issue, draw together the material, the social and lived experience of a zero-carbon development in Bangalore to explore the possibilities and complexities of transformational change. In so doing they highlight that climate change is an important urban issue – large populations in cities are vulnerable to its effects and cities themselves are significant sources of emissions. However, research on climate change in urban areas has often focused on policy for infrastructure development: water, sanitation, energy, transport, the built environment rather than the material infrastructure itself or the practices and political economies that sustain them. Bulkeley and Castán Broto (2012) instead focus a on niche or experimental project – a zero-carbon gated community in Bangalore aimed at the wealthier classes. They view this as a socio-technical experiment co-produced through the interrelation of social and material elements, rather than the product of an ideal urban policy model of evidence, goals, planning and action. Indeed, this socio-technical experiment mediates traditional governance responses and foregrounds governance itself as a contested, partial and fragmented process.

Bulkeley and Castán Broto (2012) propose a different way of thinking about urban climate governance through processes of making, maintaining and living. 'Making' relates to the processes of assembling the material and semiotic networks necessary to legitimate alternative experiments outside mainstream policy – in this case the T-Zed zero-carbon development of 16 houses and 75 apartments built from sustainable materials which offer residents high-end but low-emission lifestyles with reduced dependence on the city's resources such as water and electricity. 'Maintaining' refers to the processes of readjustment that takes place in order to deal with the experiment within the political economy and political ecology of the city – manifest in questions over water supplies, landownership disputes and the relationship between the developers and the residents. Finally, through 'Living' the lived experience and everyday practices of the residents are examined. The experiment produced new forms of conduct and normalisation but also conflict about the purpose and future of the housing development as a climate change experiment.

Bulkeley and Castán Broto conclude that the T-Zed development has become part of an emergent low-carbon urbanism by: first, creating the space for innovation; second, reconfiguring the city's infrastructure systems; and third providing an arena in which new discourses of responsibility and carbon control for middle class residents have flourished by showing that low-carbon living is compatible with modern urbanism in Bangalore. In making this argument, they are developing a thread in contemporary urban studies across a range of disciplines which tries to understand new political orders being forged in the face of climate change, and new spaces for politics engendered therein (Braun, 2014). Braun, for example, argues that combined city growth and climate change are producing new political orders that are increasingly urban in focus and which attempt to produce strategies to manage complex relations between the city and the globe, including the city as a contributor to global climate change and a recipient of its impacts.

Bulkeley and Castán Broto and Braun's research goes some way to acknowledging that 'the surprisingly uncontrollable, surprisingly living system that is daily life' (Shove, 2012, p. 12) not only modifies extant systems of governance but insists on expanding our conceptual repertoire in order to understand it. Bulkeley and Castán Broto's research steps outside dominant approaches to climate change policy in urban areas to focus on the socio-technical relations that produce the practices of everyday life, and in so doing mediate both policy and governance. The relationship between policy and practice speaks to one of the most vexed questions for social scientists: how does change happen? Shove (2012) argues that the task of sociology, anthropology, material cultural studies and (I would add) geography is to understand how social arrangements come to be as they are. She suggests that this is important for theories of social-technical change, transition and practice, but that few of these intellectual resources have found their way into policy-making for behavioural change. Like Gatersleben et al. (2012), Shove (2012) voices disquiet about, first, the focus on the individual (in research and policy) at the expense of understanding social relations in place and, second, explanations which propose a unproblematic, linear, causative link between attitudes and behaviours. Shove's solution (2012) is to focus on 'practice'.

Practice theory takes social action to be constructed, situated and performed but, crucially, also asks how practices emerge, persist and disappear. Practices in this sense are entities 'that exist across time and space, that depend on inherently provisional integrations of elements, and that are enacted by cohorts of more and less consistent or faithful carriers' (Shove, 2012, p. 418). Thus, various resource-intensive practices that contribute to climate change – such as a daily commute by car – are not simply an individual act but have been constituted and enabled by constant repetition and are the product of historical conditions and contemporary meanings, competencies and materials. This is why they are so difficult to change with conventional policies which tend to work on the deficit model – that is, if only people had enough of the right sort of information, they would change their behaviours. Pidgeon and Fischhoff (2012, p. 35), for example, argue that

> few citizens or political leaders understand the underlying science well enough to evaluate climate-related proposals and controversies. As a result, it is hard for political leaders to generate and sustain broad public support for ambitious climate policies or for citizens to take effective personal action.

Such an approach is itself a product of the entrenched view of humans as rational agents, which I discussed above.

Our dependence in the west on a high-carbon, centralised energy system which currently provides enough power to meet our many uses is one issue which lends itself to the analysis of practice that Shrove proposes. People's energy use is a historically specific outcome of – amongst other things – the abundance of cheap energy, the invention and rapid adoption of a plethora of technology from tumble dryers to smart phones, and expectations about connectivity, information gathering, social networking and communication. This is why changing elements of our practices of energy consumption is not straightforward. Moving to a decentralised energy (DE) system is frequently proposed by advocates and academics as a means of reducing emissions but the means by which this transition can be achieved are still opaque. Wiersma and Devine-Wright (2014, p. 456) seek to go beyond a functional definition of DE as 'the supply of electricity and heat generated on or near the site where it is used' using renewable energy sources like solar, hydro, wind and biomass, to understand it as a 'multi-dimensional concept, encompassing technical, financial, political and behavioural aspects'. A wide range of public, private, third sector and community organisations are involved in developing DE projects in urban areas, with different motivations, drivers and levels of success. The research shows that the DE sector is highly heterogenous, and the success of any project is dependent on the participants' skills, access to resources, incentives, governance and more esoteric factors like local structures of feeling, trust and altruism. Such findings obviously present policy-makers with a challenge as one-size-fits-all approaches to encouraging the development of DE along with other attempts to shift behaviours will clearly have to be rethought.

The questions of regulation and incentivisation of DE raised by Wiersma and Devine-Wright raises the broader issue of whether emissions can be accounted for and ultimately reduced through the use of complex financial instruments and accounting procedures. Such issues go well beyond a concern with the individual and their behaviours and instead focus on the way emissions are produced by production, supply chains, trade and other functions of the global economy. Turner, Xin Cui, Jung Ha, and Hewings (2012) offer a detailed mathematical evaluation of whether input–output accounting techniques are appropriate for tracking pollution embodied in complex economic interactions and supply chains. They are especially concerned with the scale at which such accounting methods are applied to sub-national regional economies.

While the mathematics of such research is beyond the understanding of most ordinary mortals, such research raises important broader questions. First, this research demonstrates that scale – both temporal and spatial – is of utmost importance. When Stern (2006) called climate change the widest ranging market failure ever seen, he identified the solution in the instruments of the market itself. Key policy instruments would operate to mitigate climate change, such as 'taxes, trading based on the allocations of property rights, and regulation' (p. 7), operating at the scale of the state, or federated states such as the EU. These property rights could be allocated 'over different time horizons and across countries, firms and individuals in different ways'. At the global scale, aggregate emissions targets would determine the sum of rights to emit.

Second, markets, economics and growth are not a-political. The economic question that Turner et al. (2012) ask about the efficacy of input–output accountancy methods is also fundamentally a political one, about the distribution of the cost of climate change, the apportionment of responsibility for pollutants, and the negotiation between region and nation over economic winners and losers. Thus political capital is also expended when economic policies for climate change are implemented – as in the case of Australia's recent experience with carbon tax (Robson, 2014). Third, Turner et al's (2012) concern with the degree to which consumers or producers share the responsibility for pollution demonstrates, if only implicitly, how indifferent most of us are to the global processes that bind us to distant makers, exporters and providers (Cook & Woodyer, 2012). As Turner et al. observe, human consumption decisions lie at the heart of the climate change problem, at whatever scale.

Conclusion

For Law and Urry (2004, p. 390), the work of social science is to 'interfere in the realities of [the] world, to make a difference, to engage in ontological politics, and to help shape new realities'. There is no doubt that the cogs of the social science research machine are now grinding across a number of disciplines where interest in climate change is burgeoning. And, like the mill of the gods, these cogs are grinding very small, producing detailed empirical work on a wide range of topics including behaviour, identity, values, economics, policy, governance, regulation, everyday life, co-production – the list goes on. Meanwhile, and perhaps paradoxically, the relationship between the natural sciences and the social sciences is being managed through a series of meta-narratives of the social sciences which seek to bring consistency and order to the monstrous anthill on the plain (to misquote Wordsworth).

Current attempts to assert the value of social science are political, and do not really leave much room for reflexive, critical practice. The prescription for a successful engagement between the natural and social sciences is for each to better understand the other, but in order to do this we must first better understand ourselves. Questions of ontology and epistemology are rarely brought to the fore. Methodological assumptions and interpretative frameworks are described but rarely interrogated. The structure of academic publishing regimes, funding and career trajectories in which one seeks to thrive in one's own disciplinary area tend to mitigate against taking the time to reflect on one's own scholarly practice.

Ultimately, however, much work is still required to destabilise taken-for-granted assumptions about: (i) the nature of climate change, paying attention to its complex ontology and the knowledge-making practices that have brought it into view, as well as its likely manifestations, unevenly distributed in time and space and differently mitigated by relative levels of resilience; and (ii) academic knowledge as a privileged site in which climate change can be apprehended at the expense of other ways of knowing, doing and being in the world. In this paper I have organised some extant work from disparate social science disciplines around spaces, sites and practices as a way of demonstrating not only that apparently different scholarly accounts have at least some congruence but that grounding our research in the places and practices of everyday life is productive. Needless to say, much still needs to be done. But most crucially we should learn to be more reflexive and embrace the challenge to our epistemic assumptions if we are to allow the social sciences to address the complex problem of climate change.

Acknowledgements

I would like to acknowledge Michael Leyshon for help in fine tuning the manuscript.

Note

1. Throughout this paper I refer to the social sciences in the plural but note that recent reports (Hackmann & StClair, 2012; ISSC/UNESCO, 2013) refer to social science in the singular, which suggests a theoretical, epistemological, methodological and empirical consistency which is counterproductive to the claims for the usefulness of the social sciences in addressing climate change.

References

Brace, C., & Geoghegan, H. (2011). Human geographies of climate change: Landscape, temporality, and lay knowledges. *Progress in Human Geography*, *35*, 284–302. doi:10.1177/0309132510376259

Braun, B. P. (2014). A new urban dispositif? Governing life in an age of climate change. *Environment and Planning D: Society and Space*, *32*, 49–64. doi:10.1068/d4313

Bulkeley, H., & Castán Broto, V. (2012). Urban experiments and climate change: Securing zero carbon development in Bangalore. *Contemporary Social Science*, *9*, 393–414. doi:10.1080/21582041.2012.692483

Castree, N., Adams, W. M., Barry, J., Brockington, D., Büscher, B., Corbera, E., ... Wynne, B. (2014). Changing the intellectual climate. *Nature Climate Change*, *4*, 763–768. doi:10.1038/nclimate2339

Cook, I. J., Woodyer, T. (2012). Lives of things. In E. Sheppard, T. Barnes, & J. Peck (Eds.), *Wiley-Blackwell companion to economic geography* (pp. 226–241). Oxford: Wiley-Blackwell.

Esbjörn-Hargens, S. (2010). An ontology of climate change. *Integral Pluralism and the Enactment of Multiple Objects Journal of Integral Theory and Practice*, *5*, 43–174.

Funtowicz, S. O., & Ravetz, J. R. (1993). Science for the post-normal age. *Futures*, *25*, 739–755. doi:10.1016/0016-3287(93)90022-L

Gabrys, J., & Yusoff, K. (2012). Arts, sciences and climate change: Practices and politics at the threshold. *Science as Culture*, *21*, 1–24. doi:10.1080/09505431.2010.550139

Gatersleben, B., Murtagh, N., & Abrahamse, W. (2012). Values, identity and pro-environmental behavior. *Contemporary Social Science*, *9*, 374–392. doi:10.1080/21582041.2012.682086

Geoghegan, H., & Leyshon, C. (2012). Erratum to: On climate change and cultural geography: Farming on the Lizard Peninsula, Cornwall, UK. *Climatic Change*, *113*, 55–67. doi:10.1007/s10584-012-0417-5

Geoghegan, H., & Leyson, C. (2012). On climate change and cultural geography: Farming on the Lizard Peninsula, Cornwall, UK. *Climatic Change*, *113*, 55–66. doi:10.1007/s10584-012-0417-5

Goodman, D. (2001). Ontology matters: The relational materiality of nature and agro-food studies. *Sociologia Ruralis*, *41*, 182–200. doi:10.1111/1467-9523.00177

Gregory, D., Johnston, R., Pratt, G., Watts, W., & Whatmore, S. (2009). *The dictionary of human geography* (5th ed.). London: Wiley Blackwell.

Hackmann, H., & St Clair, A. L. (2012). *Transformative cornerstones of social science research for global change*. Paris: ISSC.

Hall, J., & Pidgeon, N. (2010). A systems view of climate change. *Civil Engineering and Environmental Systems*, *27*, 243–253. doi:10.1080/10286608.2010.482659

Hulme, M., Dessai, S., Lorenzoni, I., & Nelson, D. R. (2009). Unstable climates: Exploring the statistical and social constructions of 'normal science'. *Geoforum*, *40*, 197–206. doi:10.1016/j.geoforum.2008.09.010

ISSC/UNESCO. (2013). *World social science report 2013: Changing global environments*. Paris: OECD and UNESCO.

Jaspal, R., Nerlich, B., & Cinnirella, M. (2014). Human responses to climate change: Social representation, identity and socio-psychological action. *Environmental Communication: A Journal of Nature and Culture*, *8*, 110–130. doi:10.1080/17524032.2013.846270

Kobayashi, A., & Mackenzie, S. (2014). *Remaking Human Geography*. London: Routledge.

Knapp, C. N., & Trainor, S. F. (2013). Adapting science to a warming world. *Global Environmental Change, 23*, 1296–1306. doi:10.1016/j.gloenvcha.2013.07.007

Law, J., & Urry, J. (2004). Enacting the social. *Economy and Society, 33*, 390–410. doi:10.1080/0308514042000225716

Leyshon, C. (2014). Cultural ecosystem services and the challenge for cultural geography. *Geography Compass, 8*, 701–709.

Leyshon (née Brace), C., & Geoghegan, H. (2012). Anticipatory objects and uncertain imminence: Cattle grids, landscape and the presencing of climate change on the Lizard Peninsula, UK. *Area, 44*, 237–244. doi:10.1111/j.1475-4762.2012.01082.x

Liu, J., Dietz, T., Carpenter, S. R., Alberti, M., Folke, C., Moran, E., … Provencher, W. (2007). Coupled human and natural systems. *Ambio: A Journal of the Human Environment, 36*, 639–649.

Lowe, P., Phillipson, J., & Wilkinson, K. (2013). Why social scientists should engage with natural scientists. *Contemporary Social Science: Journal of the Academy of Social Sciences, 8*, 207–222. doi:10.1080/21582041.2013.769617

McCarthy, J., Chen, C., López-Carr, D., & Endemanõ Walker, B. L. (2014). Socio-cultural dimensions of climate change: Charting the terrain. *GeoJournal.* doi:10.1007/s10708-014-9546-x

Moon, K., & Blackman, D. (2014). A guide to understanding social science research for natural scientists. *Conservation Biology, 28*(5), 1167–1177. doi:10.1111/cobi.12326

Nagel, J. (2012). Intersecting identities and global climate change. *Identities: Global Studies in Culture and Power, 19*, 467–476. doi:10.1080/1070289X.2012.710550

Naustdalslid, J. (2011). Climate change - the challenge of translating scientific knowledge into action. *International Journal of Sustainable Development & World Ecology, 18*, 243–252. doi:10.1080/13504509.2011.572303

Oldfield, F., Barnosky, A. D., Dearing, J., Fischer-Kowalski, M., McNeill, J., Steffen, W., & Zalasiewicz, J. (2014). The Anthropocene Review: Its significance, implications and the rationale for a new transdisciplinary journal. *The Anthropocene Review, 1*, 3–7.

Paschen, J. A., & Ison, R. (2014). Narrative research in climate change adaptation: Exploring a complementary paradigm for research and governance. *Research Policy, 43*, 1083–1092. doi:10.1016/j.respol.2013.12.006

Pidgeon, N., & Fischhoff, B. (2011). The role of social and decision sciences in communicating uncertain climate risks. *Nature Climate Change, 1*, 35–41. doi:10.1038/nclimate1080

Popa, F., Guillermin, M., & Dedeurwaerdere, T. (2014). A pragmatist approach to transdisciplinarity in sustainability research: From complex systems theory to reflexive science. *Futures,* doi:10.1016/j.futures.2014.02.002

Robson, A. (2014). Australia's carbon tax: An economic evaluation. *Economic Affairs, 34*, 35–45. doi:10.1111/ecaf.12061

Shove, E. (2012). Putting practice into policy: Reconfiguring questions of consumption and climate change. *Contemporary Social Science, 9*, 415–429. doi:10.1080/21582041.2012.692484

Stern, N. (2006). What is the economics of climate change? *World Economics, 7*, 1–10. Retrieved at: http://www.world-economics-journal.com/What%20is%20the%20Economics%20of%20Climate%20Change.details?AID=238

Thiry, P. H. (1770). The System of Nature or, the Laws of the Moral and Physical World (Système de la Nature ou Des Loix du Monde Physique et du Monde Moral). Retrieved from: http://www.gutenberg.org/ebooks/8909

Turner, K., Xin Cui, C., Jung Ha, S., & Hewings, G. (2012). Input–output analyses of the pollution content of intra- and inter-national trade flows. *Contemporary Social Science, 9*, 430–455. doi:10.1080/21582041.2012.692808

Weaver, C. P., Mooney, S., Allen, D., Beller-Simms, N., Fish, T., Grambsch, A. E., … Winthrop, R. (2014). From global change science to action with social sciences. *Nature Climate Change, 4*, 656–659. doi:10.1038/nclimate2319

Whatmore, S. (2006). Materialist returns: Practising cultural geography in and for a more-than-human world. *Cultural Geographies, 13*, 600–609. doi:10.1191/1474474006cgj377oa

Wiersma, B., & Divine-Wright, P. (2014). Decentralising energy: Comparing the drivers and influencers of projects led by public, private, community and third sector actors. *Contemporary Social Science, 9*, 456–470. doi:10.1080/21582041.2014.981757

Yusoff, K., & Gabrys, J. (2011). Climate change and the imagination. *WIRES Climate Change, 2*, 516–534. doi:10.1002/wcc.117

Values, identity and pro-environmental behaviour

Birgitta Gatersleben[a], Niamh Murtagh[a] and Wokje Abrahamse[b]

[a]Department of Psychology, University of Surrey, Guildford, UK; [b]University of Victoria, Canada

The importance of understanding and promoting pro-environmental behaviour among individual consumers in modern Western Societies is generally accepted. Attitudes and attitude change are often examined to help reach this goal. But although attitudes are relatively good predictors of behaviour and are relatively easy to change they only help explain specific behaviours. More stable individual factors such as values and identities may affect a wider range of behaviours. In particular factors which are important to the self are likely to influence behaviour across contexts and situations. This paper examines the role of values and identities in explaining individual pro-environmental behaviours. Secondary analyses were conducted on data from three studies on UK residents, with a total of 2694 participants. Values and identities were good predictors of pro-environmental behaviour in each study and identities explain pro-environmental behaviours over and above specific attitudes. The link between values and behaviours was fully mediated by identities in two studies and partially mediated in one study supporting the idea that identities may be broader concepts which incorporate values. The findings lend support for the concept of identity campaigning to promote sustainable behaviour. Moreover, it suggests fruitful future research directions which should explore the development and maintenance of identities.

Introduction

Modern societies place a high value on economic prosperity. Individuals who live in these societies are continuously exposed to cultural values which promote the acquisition of wealth and material possessions. But there is increasing concern about the environmental damage engendered by current levels of consumerism (Jackson, 2009). It is therefore vital to promote pro-environmental behaviour and reduce consumption. Within the area of psychology a significant amount of research has been conducted to understand the variables that affect pro-environmental behaviours.

Much of this research focuses on the Theory of Planned Behaviour (TPB; Ajzen & Fishbein, 1974) and the Norm Activation Model (NAM; Schwartz, 1977). It is worth noting that these models aim to explain intentional or planned behaviour and may not be suitable for explaining habitual behaviour (Steg & Vlek, 2009). The TPB suggests that pro-environmental behaviour is more likely to occur when people have a positive attitude towards such behaviour, believe significant others already do it (perceived descriptive social norm) or believe it should be done (perceived injunctive social norm) and when they feel they can adopt the behaviour (perceived behaviour control). The NAM suggests that altruistic behaviour (and therefore also pro-environmental behaviour according to some) is more likely when people feel a sense of moral obligation to adopt such behaviour. Moral obligation is a function of awareness of the consequences of the behaviour for others and a sense of personal responsibility. There is now plenty of support for these models (for overviews see Bamberg & Möser, 2007; Steg & Vlek, 2009). Moreover, it has been shown that the variables in these models are affected by general and pro-environmental values (e.g. Schultz & Zelezny, 1999; Oreg & Katz-Gero, 2006; Groot & Steg, 2007) and environmental identities (Stern & Dietz, 1994; Nigbur *et al.*, 2010; Whitmarsh & O'Neill, 2010). Values and identities, however, are rarely studied together and we therefore know little about their relationship and relative impact on behaviour.

Many behaviour change interventions focus on attitudes. A person's attitude towards pro-environmental behaviour can be a good predictor of such behaviour (see Staats, 2003). Attitudes are relatively easy to change and can alter with new information or circumstances (Ajzen, 2005). But attitudes tend to be measured with respect to a specific target object or event and are therefore relatively narrow. An attitude towards one behaviour may not necessarily be related to another behaviour. For instance, people who have a positive attitude towards recycling are more likely to recycle, but this does not mean they also cycle to work or use ecological washing powder. Similarly pro-environmental behaviour in one domain does not necessarily correlate strongly with pro-environmental behaviour in another (e.g. Karp, 1996; Corraliza & Berenguer, 2000; Milfont *et al.*, 2006; Oreg & Katz-Gerro, 2006; Dolnicar & Grun, 2009) and engagement in one pro-environmental behaviour does not necessarily spillover to another (Thøgersen & Ölander, 2003). Yet there is also evidence of some consistency in individuals' behaviour (Thøgersen, 2004). Thøgersen (2004) suggests that spillover can occur, but it is more likely in behaviours that are conceptually similar (e.g. recycling glass or paper) than in behaviours which are very dissimilar (e.g. recycling glass and cycling to work). Indeed some go as far as to suggest that pro-environmental behaviour can in fact be perceived as a uni-dimensional rather than a multi-dimensional concept because such behaviours are linked through a common goal—protecting the environment (e.g. Kaiser & Wilson, 2004).

It seems valuable to examine the relative importance people attach to more general goals such as protecting the environment if this helps understand pro-environmental behaviour across different contexts. There is evidence to suggest that people who behave more pro-environmentally across contexts rate particular values highly (Thøgersen & Ölander, 2003) and that pro-environmental behaviours are influenced by such values (Schwartz & Bilsky, 1990; Lindenberg & Steg, 2007). Another stable concept that has been studied in this context is that of self-identity. Again there is evidence to suggest that different consumer behaviours are related to the extent to which people perceive themselves as a typical person who would adopt such a behaviour (e.g, Sparks & Shepherd, 1992; Whitmarsh & O'Neill, 2010). Although the relative importance or salience of identities are to an extent context dependent (e.g. at work, being a researcher is more important to me than being a mother), values and identities are generally stable factors that transcend specific situations. The extent to which you see yourself as an environmentally friendly person, for instance, is likely to be related to a wide range of pro-environmental behaviours including waste, transport and buying behaviours. These factors may operate to guide behaviours in multiple situations and thus offer broader ranging insights into determinants of 'green' behaviour. Indeed, some have argued that understanding and leveraging more fundamental aspects of the person such as values and identity is critical in moving towards sustainable behaviours (www.identity campaigning.org). Unless these deeper constructs are engaged, any change towards pro-environmental behaviour will be piecemeal, slow and disjointed, with each behaviour adopted or rejected separately by individuals, with the risk of 'rebound' ('greener' behaviour in one domain leading to less sustainable behaviour in another) undermining any gains (Crompton & Kasser, 2010; Druckman *et al.*, 2011).

There is significant evidence that values and identities play a role in explaining and predicting pro-environmental behaviour. However, very few studies have looked at values and identities simultaneously and we know little, therefore, about the relative importance of each of these constructs in understanding pro-environmental behaviour.

Self-identity

Self-identity refers to how an individual sees him/herself, and can encompass all aspects of the self such as physical attributes, preferences, values, personal goals, habitual behaviour, personality traits and personal narratives (Pillsbury, 1934; McAdams, 1995). Individuals tend to present themselves in ways that are congruent with their self-identity (Burke & Reitzes, 1991), and this extends to behaviour (Callero, 1985; Sparks & Shepherd, 1992) including consumption (Oyserman *et al.*, 2007; Dittmar, 2010). Although identity represents an individual's subjective perspective on the self, identities are formed through social interaction. Theorists in the symbolic interactionist tradition proposed the development of the self through reflection from others in social exchanges (Mead, 1934; Breakwell, 1986) and Stets and Burke (2000) proposed that identities develop through processes of self-categorisation and identification. People thus develop multiple identities e.g. I am a woman, I am a researcher, I am an environmentalist. Multiple identities are proposed as being managed in a 'hierarchy of salience' (Stryker, 1984): identities vary in salience, and particular identities, such as gender, are likely to be chronically salient.

Identities can form barriers to pro-environmental behaviours. For instance, Stradling *et al.* (1999) found that car drivers are less willing to reduce their car use when they derive a sense of personal identity from driving. Identities can also motivate 'green' behaviour. An environmental identity reflects the extent to which people indicate that environmentalism is a central part of who they are, and a number of studies have shown that an environmental identity increases engagement in pro-environmental actions. For example, Whitmarsh and O'Neill (2010) found that people with a 'green' identity more often act pro-environmentally. Similarly, Van der Werff *et al.* (2011) found that an energy saving identity is positively related to intentions to conserve energy.

Exploring how the influence of identities on behaviour may be theoretically modelled, several studies have considered identities in conjunction with the TPB (Ajzen & Fishbein, 1974). TPB proposed that intention to perform a behaviour is predicted by three factors: attitudes (is it a good or bad thing to do?), subjective norms (what do others think I should do?) and perceived behavioural control (can I do it?). Empirical results have demonstrated that, over and above these variables, identity can explain behaviours including consumer behaviour (purchasing fashionable watches, trendy backpacks and mobile phones; Manetti *et al.*, 2002), 'green' consumption (Sparks & Shepherd, 1992) and recycling (Nigbur *et al.*, 2010). The conclusion from these studies was that the TPB should be extended to include identity as a predictor of behaviour.

Values

Values may be defined as 'concepts or beliefs, [about] desirable end states or behaviours, [which] transcend specific situations, [and] guide selection or evaluation of behaviour and events, and are ordered by relative importance' (Schwartz & Bilsky, 1990, p. 878). Schwartz (1990, 1992) developed a Values Inventory, comprising 56 'guiding principles in life' and his work has been validated in many transnational studies. This research suggests that human values can be grouped into 10 motivational domains and two dimensions (self-enhancement versus self-transcendence and openness to change versus conservatism). Using Schwartz's inventory, Stern (2000) and colleagues have suggested that three values underlie environmental concern: egoism, altruism and biospherism. De Groot and Steg (2007, 2008) further developed this idea, creating and evaluating among a wide range of samples, a short rating scale which measures these three value orientations.

There are many other measures of environmental values (see Dietz *et al.*, 2005 for an overview). The New Ecological Paradigm (NEP) is the most commonly used (Dunlap *et al.*, 2000). It measures the extent to which people have an anthropocentric versus an ecocentric worldview. NEP has been shown to relate negatively to egoism, and positively to biospherism (De Groot & Steg, 2008) and to self-transcendence (Schultz & Zelezny, 1999). Stern and colleagues posited that general values affect more specific values (NEP). NEP affects awareness of consequences (of environmentally damaging behaviours) and subsequently awareness of responsibility to reduce these consequences. This will then result into a sense of obligation to reduce the threat and therefore affect pro-environmental behaviour. Several studies have supported (parts of this) model (e.g. Schultz & Zelezny, 1999; De Groot & Steg, 2007, 2008).

A final value concept that may be relevant when studying pro-environmental consumer behaviours is materialism. Richins (2004) developed a materialistic values scale (MVS) to measure 'the importance people ascribed to the ownership and acquisition of material goods in achieving major life goals or desired states' (p. 210). Negative correlations tend to be found between materialism and environmental values (Banerjee & McKeage, 1994; Clump *et al.*, 2002; Brown & Kasser, 2005; Hirsh & Dolderman 2007; Kilbourne & Pickett, 2008; Gatersleben *et al.*, 2010). The reason why these values may be negatively related is often explained on the basis of Schwartz's work on general values (e.g. Schwartz & Bilsky, 1990). Materialism is positively related to self-enhancement (Richins, 2004; Kilbourne *et al.*, 2005) and egoism (Gatersleben *et al.*, 2010) whereas environmental values are positively related to self-transcendence (Stern & Dietz, 1994; Schultz & Zelezny, 1999).

Values, identity and behaviour

Only recent work has started to examine the role of both values and identity (e.g. Snelgar, 2003; Whitmarsh & O'Neill, 2010; Van der Werff *et al.*, 2011). We know little about the link between values and identity, although values have been seen as an integral part of identity. MacAdams (1995) conceptualised identity as an integrated life story: 'what person the person is trying to make' (p. 306). Within this narrative, values are drawn upon to explain behaviour and to characterise the self. Hitlin (2003) proposed that values form a cohesive core of personal and social identities, arguing that a values-based conception of personal identity influences the formation of a role or social identity. He showed that relevant values along the self-enhancement/self-transcendence dimension are significant predictors of the volunteer identity, controlling for previous measures of the identity.

Values are generally perceived as fairly distal determinants of behaviour which influence behaviour via more proximal determinants, such as beliefs, specific attitudes and norms (e.g. Eagly & Chaiken, 1993; Stern *et al.*, 1995). Identities, however, are broader concepts encompassing many aspects of the self, including psychological processes (including behaviours) which people may adopt for maintaining and protecting the self (Breakwell, 1986). For instance, if being environmentally friendly is an important part of who you are, recycling, voting for the green party and buying ecological products may all be important things to do in order to express, maintain and protect that identity.

We propose then that values are components, even central components, of identity. Identity is the theoretically broader construct, encompassing many other aspects of the self, such as self-image, social roles (Stryker, 1984) and psychological processes for maintaining and protecting the self (Breakwell, 1986). It can be suggested that identity may mediate the relationship between values and behaviours because values are part of one's identity: if you describe yourself as an environmentally friendly person you are likely to hold strong environmental values and behave pro-environmentally.

The current research aims to explore in more detail the relationship between values, identity and pro-environmental behaviours. The relationship between identity and two major theories of planned behaviour (TPB, Ajzen & Fishbein (1974) and NAM, Schwartz (1977)) are also investigated. Secondary analyses were conducted on three different data sets from studies among UK residents. In each of these studies, questions were included on pro-environmental behaviour, and on identity, values or both. In the analysis below, the first study examines the extent to which identity may mediate the relationship between materialistic values (MVS, Richins, 2004) and environmental values (NEP, Dunlap *et al.*, 2000) on the one hand and intentions to buy fair trade produce on the other. The second study examines the extent to which identity mediates the link between biospheric, altruistic and egoistic values (De Groot & Steg, 2007, 2008) and self-reported pro-environmental behaviour. The final study examines whether identity explains variance in intentions to adopt a range of pro-environmental behaviours, alongside variables from TPB (Ajzen & Fishbein, 1974) and NAM (Schwartz, 1977).

Study 1: values, identity and ecological purchases

A survey study was conducted among English households in 2001 to examine community engagement and attitudes and perceptions in relation to sustainable lifestyles. The survey was distributed in two areas in England, one urban and one rural area. Respondents could win a £70 voucher (just over 100 Euro or US dollar in 2001) if they returned the completed questionnaire in the freepost envelope provided. A total of 2000 surveys were sent out and 266 were returned (a 13% response rate). Just over half the respondents came from the rural area (54%) and about two-thirds were female (64%). About a third of the respondents were between 16 and 45 years old, another third was between 45 and 65 years old and the remainder were 65 or older. The average annual income of the respondents ranged from less than £10,000 to more than £100,000, with an average of around £35,000 (above the national average of around £28,500 in 2001 (ONS statistics; www.ons.gov.uk; approximately €54,000, $52,000 in 2001).

Materialism was measured with the MVS developed by Richins (2004). Scores can range from 1 to 5; the mean score was calculated for each respondent across the 15 items of this scale. The scale had a high internal consistency ($\alpha = 0.80$). Materialism was generally low ($M = 2.47$, $SD = 0.51$). It was not related to age, gender or income.

Environmental values were measured with the NEP (Dunlap *et al.*, 2000). Scores on the NEP were relatively high ($\alpha = 0.78$; $M = 3.69$, $SD = 0.52$; $1 = $ low, $5 = $ high). The NEP was not related to age, gender or income but was negatively related to materialism ($r = -0.19$, $p = 0.03$).

Pro-environmental behaviour was measured by asking respondents how often they buy Fair Trade food products and organic food products. These two items were combined into one variable by calculating their mean score ($M = 2.71$, $SD = 0.96$). Buying behaviour was positively related to income ($r = 0.29$, $p < 0.001$) but not to age or gender.

Identity participants were asked to what extent they considered themselves to be different consumer types (e.g. health conscious or frugal). Factor analyses revealed three factors explaining 54% of the variance in total. The first factor (explaining 20% of the variance) captured the extent to which respondents perceived themselves to be 'hedonist consumers' (fashion conscious, reckless, self-indulgent, compulsive and not cautious). The second factor (explaining 20% of the variance) captured the extent to which respondents perceived themselves to be 'conscious consumers' (health conscious, green, fitness conscious, ethical). The third factor grouped the remaining two items (eco-centric and a non-consumer). On the basis of the first two factors, two new variables were created by calculating the means over items which had factor loadings of 0.50 or above on the relevant factor in the rotated factor solution: 'hedonist consumer' ($\alpha = 0.66$; $M = 2.27$, $SD = 0.61$) and 'conscious consumer' ($\alpha = 0.66$; $M = 3.51$, $SD = 0.63$). The extent to which respondents identified as a conscious consumer was not related to age and gender. Females were more likely to identify with a hedonist consumer identity ($M = 2.35$, $SD = 0.60$) than male respondents ($M = 2.13$, $SD = 0.60$; $t = 2.80$ (259), $p = 0.006$). Moreover, income was positively related to identifying with a conscious consumer identity ($r = 0.17$, $p = 0.007$) as well as a hedonist consumer identity ($r = 0.17$, $p = 0.008$).

Results

Simple correlations were computed to examine the link between identities and values. Materialistic values (MVS), but not the NEP, were positively related to a hedonist consumer identity ($r = 0.28$, $p < 0.001$). The extent to which respondents saw themselves as conscious consumers was positively related to NEP ($r = 0.16$, $p = 0.10$) but not to MVS.

Regression analyses were conducted to examine whether values are related to pro-environmental behaviours. The NEP was positively related to pro-environmental behaviour and materialism was not significantly related (Step 1, Table 1). When identities were included in the regression (Step 2), significantly more variance was explained ($\Delta R^2 = 0.20$, $F(2, 254) = 34.90$, $p < 0.001$). Both hedonist and conscious consumer identities were related to pro-environmental behaviour. When identities were included, the relationship between environmental values and behaviours was weaker, suggesting that identities mediate the link between values and behaviours. To test this a Sobel mediation test was conducted (Baron & Kenny, 1986). The Sobel test for the conscious consumer identity was significant ($z = 2.28$, $p = 0.01$), showing that this identity mediated the relationship between NEP and pro-environmental behaviour. Mediation was partial with a small significant relationship remaining when identity was included.

Table 1. Regression of pro-environmental behaviour onto MVS, NEP and identities

	Step 1			Step 2		
	$\Delta R^2 = 0.04$; $F(2, 257) = 6.94$***			$\Delta R^2 = 0.24$; $F(4, 254) = 21.59$***		
	B	Error B	β	B	Error B	β
(Constant)	3.53	0.57		5.74	0.57	
MVS	−0.17	0.12	−0.09	−0.23	0.11	−0.12*
NEP	0.34	0.12	0.18**	0.21	0.10	0.12*
ID hedonist				0.26	0.09	0.17**
ID conscious				0.63	0.08	0.42***

Note: Multicollinearity between identities was not detected. *$p < 0.05$; **$p < 0.01$; ***$p < 0.001$.

Study 2: identity, values and the New Environmental Paradigm

A survey was send to a random sample of households in two areas in the UK, one city in the North and one town in the South. The study examined the role of values and identity in explaining different pro-environmental behaviour. One thousand question-naires were sent out in 2009, and 135 were returned (a response rate of 13.5%), of which 36% were from the North. Just under half of the respondents were female (47%). About a third of the respondents were under 50, another third were between 50 and 70 years of age and another third were over 70. Two-thirds earned over £25,000 with a third of the sample earning more than £50,000 per annum (Average household income in 2009 approximately £36,000 (ONS statistics; www. ons.gov.uk); approximately €36,000, $52,000 in 2009).

Identity: three questions were asked for four different consumer identities (health conscious, environmentally friendly, moral and frugal). These items included ques-tions such as 'Being ... is an important part of who I am' (1 = strongly agree, 5 = strongly disagree). The questions were based on previous research (e.g. Sparks & Shepherd, 1992; Hinds & Sparks, 2008) and a qualitative study (Evans & Abrahamse, 2010). For each identity, a scale was computed across the three relevant items. The health ($M = 4.03$, $SD = 0.68$) and environmental identity ($M = 3.50$, $SD = 0.85$) scale showed very good reliability (Cronbach $\alpha > 0.80$), the moral identity scale showed good reliability ($\alpha = 0.70$; $M = 3.78$, $SD = 0.66$) but the frugal identity scale showed very poor reliability (0.25) which could be improved significantly ($\alpha = 0.84$; $M = 3.96$, $SD = 0.82$) upon removal of one item ('As a person it is important to me that I attempt not to be wasteful'). This may be because of the wording of the question, which includes a double negative. Such questions are more difficult to answer and this may have resulted in increased random error in responses.

A higher income was negatively related to environmental ($r = -0.29$, $p = 001$), moral ($r = -0.35$, $p < 0.001$) and frugal identities ($r = -0.27$, $p = 0.003$). Older people in the sample were more frugal ($r = 0.20$, $p = 0.02$). Women were more likely to identify with an environmental consumer identity ($M = 3.71$, $SD = 0.77$) than men ($M = 3.32$, $SD = 0.88$; $t = 2.69$ (131), $p = 0.008$). They were also more likely to identify with a moral consumer identity ($M = 3.92$, $SD = 0.59$) than men did ($M = 3.66$, $SD = 0.70$; $t = 2.24$ (131), $p = 0.027$).

Values: values were measured using the values scale developed by De Groot and Steg (2008). Respondents were asked to indicate how important 13 different values were as a guiding principle in their lives (-1 'goes against my principles', 0 'not important' to 7 'extremely important'). Cronbach alpha for the 5 egoistic values (authority, wealth, power, being influential, being ambitious) was 0.71 ($M = 2.55$, $SD = 1.31$); for the 4 altruistic values (social justice, equality, peace, being helpful) was 0.75 ($M = 5.25$, $SD = 1.20$); and for the 4 biospheric values (preventing pollution, protecting the environment, respecting the earth, unity with nature), the alpha coefficient was 0.89 ($M = 5.07$, $SD = 1.44$). Values were not related to age, gender or income.

New Environmental Paradigm: as in Study 1, respondents were asked to complete the NEP (Dunlap *et al.*, 2000). Scores on this scale were high ($\alpha = 0.81$; $M = 3.51$, $SD = 0.51$; 1 = low, 5 = high). NEP scores were not related to age, gender or income.

Pro-environmental behaviour: respondents indicated how often they adopted 20 pro-environmental behaviours, on a 5-point scale (1 = never, 5 = always). These included energy behaviours (e.g. lowering thermostat) as well as recycling, food and transport behaviours. One scale was computed on the basis of these questions and showed good reliability ($\alpha = 0.83$; $M = 3.44$, $SD = 0.50$). Those with a higher income were less likely to adopt pro-environmental behaviours ($r = -0.23$, $p = 0.008$). Women were more likely to adopt pro-environmental behaviours ($M = 3.66$, $SD = 0.42$) than men ($M = 3.25$, $SD = 0.48$; $t = 5.24$ (133), $p < 0.001$).

Results

Simple correlations explored the relationship between values and identities. Moderate to strong relationships were found (see Table 2). Both biospheric values and NEP were strongly related to environmental identity, as well as to moral and frugal identities. This suggests there may be overlap between the value and identity concepts, especially where they share related goals, such as environmental conservation or morality. Interestingly, and perhaps surprising, egoistic values were also positively related

Table 2. Correlations between values and identities

		Identity			
		Health	Environment	Moral	Frugal
Values	Biospheric	0.37***	0.68***	0.52***	0.42***
	Egoistic	0.32***	0.20*	0.30**	0.10
	Altruistic	0.34***	0.46***	0.57***	0.38***
	NEP	0.09	0.48***	0.27**	0.27**

Note: *p < 0.05, **p < 0.01, ***p < 0.001.

to health, environmental and moral identities. A positive correlation was found between egoistic and altruistic values ($r = 0.25$, $p = 0.004$). The correlation between egoistic and biospheric values was not significant ($r = 0.13$, ns). This is not in line with the literature which suggests that these values should be inversely related. It is most likely a response artefact where some people were simply more likely to agree with all questions on the scale. Expected correlations, however, were significantly higher than these unexpected correlations and supported our hypotheses.

Regression analyses were conducted to investigate relative contributions to variance in pro-environmental behaviours. A step-wise regression was carried out, with values included in the first step, and four identities added to the equation in the second step. Table 3 presents the results. In Step 1, a biospheric value was the only significant predictor. Step 2, which includes identities, explained significantly more variance (18%; $\Delta R^2 = 0.18$, $F(4, 121) = 10.38$, $p < 0.001$). Two identities contributed significant variance—environmental and frugal identities—and biospheric values become non-significant. Sobel tests showed that biospheric values were fully mediated by environmental identity (B becomes non-significant; $z = 4.65$, $p < 0.001$) and partially mediated by frugal identity ($p < 0.001$; $z = 3.46$, $p < 0.001$).

As in Study 1, NEP was significantly related to pro-environmental behaviour (see Table 4). Adding identities to this simple regression explained an additional 36% of the variance in reported behaviours ($\Delta R^2 = 0.36$; $F(4, 127) = 21.98$, $p < 0.001$). Sobel tests showed that the link between NEP and pro-environmental behaviour was fully mediated by environmental identity ($B = 0.06$, error $B = 0.07$, $p = 0.411$; $z = 5.03$, $p < 0.001$) and partially mediated by frugal identity ($B = 0.30$, error $B = 0.05$, $p = 0.002$; $z = 2.73$, $p < 0.001$).

Table 3. Regression of pro-environmental behaviour onto values and identity

	Step 1			Step 2		
	$\Delta R^2 = 0.26$; $F(3, 127) = 16.05^{***}$			$\Delta R^2 = 0.43$; $F(7, 121) = 14.90^{***}$		
	B	SE	β	B	SE	β
(Constant)	2.38	0.17		1.61	0.24	
Value Biospheric	0.13	0.03	0.40^{***}	0.03	0.04	0.09
Value Egoistic	0.04	0.03	0.11	0.01	0.03	0.01
Value Altruistic	0.06	0.04	0.14	0.02	0.04	0.05
IDhealth				0.03	0.06	0.04
IDenvironment				0.21	0.07	0.35^{**}
IDmoral				0.02	0.07	0.03
IDfrugal				0.17	0.05	0.28^{**}

Note: Due to high correlations between independent variables, we checked for multicollinearity but found no violations of assumptions.

Table 4. Regression of pro-environmental behaviour onto NEP and identity

| | Step 1 | | | Step 2 | | |
| | $\Delta R^2 = 0.12$; $F(1, 123) = 19.52$*** | | | $\Delta R^2 = 0.47$; $F(5, 127) = 23.99$*** | | |
	B	SE	β	B	SE	β
(Constant)	2.21	0.28		1.51	0.31	
NEP	0.35	0.08	0.36***	0.06	0.07	0.06
IDhealth				0.00	0.06	0.01
IDenvironment				0.27	0.06	0.45**
IDmoral				0.06	0.06	0.07
IDfrugal				0.15	0.05	0.25**

Note: Due to high correlations between independent variables, we checked for multicollinearity but found no violations of assumptions.

Study 3: identity, attitudes, norms and perceived behavioural control

An on-line survey was developed in 2007 by a commercial marketing research company on behalf of a major media group in the UK. The survey link was advertised in a range of media owned by this group (television, radio and magazines). Potential participants were offered a chance to win a range of prizes for participating in the study. The survey consisted of nearly 600 questions, most of which focused on media use (commercial television, radio and magazines), with some final questions on pro-environmental behaviours and identity.

A total number of 2293 people participated in the survey. The majority lived in England (76%), around 5% each lived in Wales, other European countries or the USA. Around a third of the respondents were between 40 and 80 years of age, a third was between 28 and 40 and another third was between 16 and 28 years old, making the sample relatively young. Just over half of the respondents (52%) were female. Over a quarter (28%) earned less than £25,000 and 17% earned more than £50,000 per annum (average household income in 2007 approximately £34,000 (ONS statistics; www.ons.gov.uk); approximately €50,000, $66,000 in 2007).

Pro-environmental attitudes and behaviours: questions were asked about five behaviours: three with negative environmental impact (using a car for grocery shopping, using a car for travelling to work, using an aeroplane to go on holiday) and two with positive impact (buying Fair Trade coffee or tea and recycling household waste). All scales had 5 scale points. For each of these behaviours, respondents were asked one question on *intention* ('To what extent do you intend to... the next time you...'; 1 = definitely not, 5 = definitely), and one question on *perceived behavioural control* ('How easy is it for you to...'; 1 = very difficult, 5 = very easy). Both of these questions were phrased with respect to sustainable behaviours, for example, 'To what extent do you intend to avoid using your car the next time you travel to work?' One question for each behaviour was asked on *attitude* ('What is your attitude towards...', 1 = Strongly disapprove, 5 = Strongly approve), one question on *(injunctive) subjective norm* ('What is the attitude of your friends towards...', 1 = Strongly disapprove, 5 = Strongly approve), and *personal norm* ('I feel guilty when I...', 1 = Strongly disagree, 5 = Strongly agree). These three questions were phrased with respect to non-sustainable behaviours. Table 5 presents means and standard deviations. For clarity, the table depicts all variables with respect to sustainable behaviour, and reverses the scores on attitudes and subjective norms.

Only very weak relationships were found with age, gender or income. All significant correlations were low (one of 0.15, and the remainder below 0.10). Of note is that there appears to be a generally linear progression for all variables across the behaviours as shown in Table 5. That is, intention, attitude, perceived behavioural control, subjective and personal norms for buying Fair Trade products were stronger than for avoiding flying on holiday, which in turn was stronger than the avoidance of car use for shopping. Repeated measures analyses showed significant linear increases in intentions ($F(1, 1654) = 3287.89$, $p < 0.001$), attitudes ($F(1, 2257) = 3257.90$, $p < 0.001$), subjective norm ($F(1, 1646) = 1434.66$, $p < 0.001$), personal norms ($F(1, 1656) = 2170.97$, $p < 0.001$) and perceived behavioural control ($F(1, 1505) = 3048.88$, $p < 0.001$) in the order in which the variables are presented in Table 5.

Identity was measured with four items asking respondents to what extent they agreed that they were a health conscious consumer ($M = 3.57$, $SD = 0.87$), a price conscious consumer ($M = 3.94$, $SD = 0.84$), an environmentally friendly consumer ($M = 3.37$, $SD = 0.83$) and a frugal consumer ($M = 3.22$, $SD = 0.86$; 5-point scale anchored at 1 = Strong disagree, 5 = Strong agree). These single item measures were analysed separately. Only very weak correlations were found between identities and demographic variables (all correlations were below 0.15, most below 0.10).

Table 5. Intentions, attitudes, perceived behavioural control, subjective and personal norms in relation to five sustainable behaviours

		Avoid car use for major grocery shop	Avoid car use for work	Not flying to holiday destination	Buying Fair Trade coffee and tea	Recycling
Intention	M	1.93	2.38	2.47	2.90	4.43
	SD	(1.21)	(1.64)	(1.26)	(1.10)	(1.04)
Attitude[a]	M	2.53	2.83	2.62	3.00	4.23
	SD	(0.86)	(0.92)	(0.86)	(0.76)	(0.99)
PBC	M	1.93	2.35	2.65	3.72	4.07
	SD	(1.17)	(1.54)	(1.23)	(1.07)	(1.19)
Subjective norms[a]	M	2.43	2.53	2.45	2.95	3.63
	SD	(0.83)	(0.84)	(0.80)	(0.65)	(0.98)
Personal norms	M	2.18	2.38	2.33	2.69	3.83
	SD	(1.01)	(1.11)	(1.04)	(1.05)	(1.16)

[a]Score reversed.

Results

Simple correlations suggested that there were small significant relationships between identities and attitude, perceived behavioural control, subjective norm and personal norms. The strongest links were found between identities and personal norms and in particular for an environmental identity (Table 6).

For each of the five behaviours, we conducted a stepwise regression (see Table 7): in the first step, intention towards pro-environmental behaviour was regressed onto TPB and NAM variables; in the second step, identities were added to the equation. For car use to work and for shopping, TPB and NAM variables appeared to be good predictors of intentions and identities did not explain additional variance. For reducing holiday flights, buying Fair Trade and recycling, however, we found a significant contribution of identities. In particular, environmental identity explained additional variance in each case for each of the behaviours over and above the TPB and NAM variables with 1% ($\Delta R^2 = 0.005$, $F(4, 1851) = 4.01$, $p = 0.003$) added for not taking holiday flights, 3% for buying Fair Trade ($\Delta R^2 = 0.028$, $F(4, 1608) = 18.81$, $p < 0.001$) and 1% for recycling ($\Delta R^2 = 0.011$, $F(4, 1862) = 10.46$, $p < 0.001$).

Table 6. Correlations between identities and attitudes, perceived behavioural control, subjective and personal norms

		Identities			
		Health	Price	Environmental	Frugal
Attitude	Car use shop	−0.05*	0.03	−0.10**	−0.05*
	Car use work	−0.07**	0.01	−0.13**	−0.04*
	Fly holiday	−0.08**	−0.04*	−0.18**	−0.09**
	Buy fair trade	−0.04*	0.00	−0.11**	−0.04*
	Recycle	−0.17**	−0.06**	−0.28**	−0.06**
Perceived behavioural control	Car use shop	−0.01	−0.08**	0.00	0.00
	Car use work	0.00	0.00	0.03	−0.01
	Fly holiday	−0.03	0.02	0.05*	0.07**
	Buy fair trade	0.13**	−0.03	0.12**	0.01
	Recycle	0.10**	0.02	0.26**	0.03
Subjective norms	Car use shop	−0.02	0.05*	−0.03	−0.02
	Car use work	−0.04	0.01	−0.07**	0.01
	Fly holiday	−0.04	−0.02	−0.06**	−0.05*
	Buy fair trade	0.01	−0.02	−0.02	−0.04
	Recycle	−0.11**	−0.05*	−0.12**	−0.06*
Personal norms	Car use shop	0.14**	0.00	0.21**	0.06**
	Car use work	0.15**	0.04	0.22**	0.06*
	Fly holiday	0.16**	0.05*	0.27**	0.11**
	Buy fair trade	0.17**	0.02	0.31**	0.05*
	Recycle	0.24**	0.13**	0.32**	0.12**

Table 7. Regression of pro-environmental intentions onto attitudes, perceived behaviour control (PBC), subjective norms, personal norms and identities

	Avoid car use for major grocery shop	Avoid car use for work	Not flying to holiday destination	Buying Fair Trade coffee and tea	Recycling
Step 1:	$\Delta R^2 = 0.54$ $F(4, 1399)$ $= 418.88***$	$\Delta R^2 = 0.61$ $F(4, 1391)$ $= 552.03***$	$\Delta R^2 = 0.45$ $F(4, 1855)$ $= 376.84***$	$\Delta R^2 = 0.38$ $F(4,1634)$ $= 249.26***$	$\Delta R^2 = 0.48$ $F(4,1894)$ $= 440.92***$
Attitude	0.17***	0.18***	0.27***	0.09***	0.17***
Subj. norms	−0.02	−0.02	0.00	0.01	0.04*
PBC	0.61***	0.66***	0.41***	0.28***	0.54***
Personal norms	0.12***	0.08***	0.18***	0.45***	0.16***
Step 2:	$\Delta R^2 = 0.54$ $F(8, 1375)$ $= 204.97***$	$\Delta R^2 = 0.61$ $F(8, 1366)$ $= 271.77***$	$\Delta R^2 = 0.45$ $F(8, 1851)$ $= 191.65***$	$\Delta R^2 = 0.41$ $F(8, 1608)$ $= 139.01***$	$\Delta R^2 = 0.49$ $F(8, 1862)$ $= 227.23***$
Attitude	0.17***	0.18***	0.27***	0.08***	0.15***
Subj. norms	−0.02	−0.02	0.00	0.02	0.04*
PBC	0.61***	0.66***	0.41***	0.27***	0.52***
Personal norms	0.11***	0.08***	0.16***	0.40***	0.13***
IDhealth	0.01	0.01	−0.02	−0.01	−0.01
IDprice	−0.03	0.03	0.02	−0.04	0.01
IDenvironment	0.02	0.00	0.06**	0.19***	0.12***
IDfrugal	0.03	0.01	0.02	−0.02	−0.01

Discussion

Secondary analyses were conducted on data from three studies. The analyses explored the relationships of identity and values on pro-environmental behaviour, and their relationship with two existing models of such behaviour: the TPB (Ajzen & Fishbein, 1974) and the NAM (Schwartz, 1977). It was hypothesised that identity would mediate the relationship between values and pro-environmental behaviour. Moreover, Study 3 examined whether identity would explain variance in intention towards pro-environmental behaviour over and above attitudes, perceived social norms, perceived behavioural control and personal norms (variables from TPB and NAM). The analyses showed full mediation by environmental identity of the relationship between biospheric values and 'green' behaviour, and between NEP and 'green behaviour' (study 2). In Study 1, a 'conscious consumer' identity was found to partially mediate the link between NEP and pro-environmental behaviour. Environmental identity was significantly related to intention to act pro-environmentally in all three studies and identities explained variance in specific pro-environmental behaviours alongside TPB and NAM variables. However, this did not hold for all pro-environmental behaviours measured. Moreover, although significant, identities appeared to contribute only a small amount of additional explanation. The hypotheses were therefore partially supported.

Although we found full mediation by environmental identity of the link between NEP and pro-environmental behaviour in Study 2, we found only partial mediation in Study 1. It is likely that this relates to the different operationalisations of the variables in the studies. Study 1 comprised only two behaviours, which specifically focused on buying Fair Trade and organic produce. These are relatively specific behaviours in that they both refer to (moral) buying behaviour. In Study 2, a wide range of different pro-environmental behaviours were combined. The variable in Study 2 may therefore have been a better reflection of general pro-environmental behaviour than the variable in Study 1 and is therefore more strongly related to environmental identities. Identities were also operationalised differently in both studies. Whereas Study 1 examined a range of identities and grouped these together into a hedonist and a conscious consumer identity, Study 2 examined more specific consumer identities. The independent and dependent variables in Study 2 therefore may have been more closely matched in operational terms. When the variables are operationalised at a similar level of specificity (e.g. general pro-environmental behaviour and environmental identities) full mediation is more likely to be found for general environmental values and for the New Environmental Paradigm.

The finding that identity is a significant predictor of intention to perform pro-environmental behaviours, alongside attitudes, subjective norms and perceived behavioural control from TPB, supports and extends previous work by Sparks and Shepherd (1992), Manetti et al. (2002), Nigbur et al. (2010) and others (see Conner & Armitage, 1998, for a review). The additional contribution of identity to intention in Study 3, however, was very small although this is in line with the findings of Conner and Armitage (1998).

Environmental identity was related to several, but not all, pro-environmental behaviours, an outcome suggested as likely by Conner and Armitage (1998). An environmental identity was related to recycling, buying Fair Trade, and avoiding flying on holiday, but not to reducing car use for work or shopping. The strongest predictor for four of the five behaviours was perceived behavioural control. So for avoiding car use, not flying to a holiday destination and recycling, intention to behave more sustainably was most strongly related to how easy the participants thought it would be. And this supports Kaiser and others who have argued that ease of action is critical (Kaiser & Wilson, 2004). Buying Fair Trade tea and coffee showed a different pattern. Personal norm was the strongest predictor. This could suggest that identities and personal norms become more important for behaviours in which the individual feels relatively free to act. In choosing consumer products, individuals may feel unconstrained and their behaviour may be guided more by how they see themselves, as 'green' or moral people for example. This could explain the findings of Sparks and Shepherd (1992) of identities contributing to 'green consumerism', and the findings of Nigbur et al. (2010), who suggested that their participants had complete freedom in choosing to recycle household waste. In contrast, where individuals feel that practical factors constrain how they act—the availability of alternatives to driving to work and going shopping, for example, these perceived constraints may dominate behaviour (see also Whitmarsh & O'Neill, 2010).

In this paper we argued that when studying pro-environmental behaviour, it is important to focus on variables which transcend specific situations but may help to promote such behaviours across a range of contexts and situations. Values and identities were presented as two such useful variables. To date most of the work on such aspects focuses on values and, in particular, environmental values such as the NEP (see Dietz et al., 2005). This paper suggests that it may be worth further exploring the role of identities. We argued that identities may encapsulate a range of psychological variables including values and the studies presented in this paper support this hypothesis. Not only did we find that environmental identities tended to explain additional variance over and above value items, we also found that the link between values and pro-environmental behaviour was either fully or partially mediated by identities, suggesting that identities explained the variance accounted for by values as well as additional variance. Drawing on the rich theoretical basis of identity may help us to understand not only how identities may influence behaviour but also how they develop and are maintained (e.g. Breakwell, 1986). This offers a particularly fruitful avenue to study in order to promote changes in behaviours and to meet the longer term goal of the development of more sustainable lifestyles.

Drawing together the two main implications from the findings: that values (and perhaps TPB and NAM variables) are aspects of self-identity, and that identities will vary in the extent to which they guide particular sustainable behaviours, suggestions for future research and the promotion of pro-environmental behaviour may be made. Our finding that mediates the relationship between values and behaviour lends weight to, and refines, the argument of 'identity campaigning': not only values but self-identity more broadly are important as predictors of 'green' behaviour.

The relative salience of different identities may play a role. Identities develop over the lifespan. We may want to explore how an understanding of identity development could inform promotion of sustainable behaviour. Moreover, more work is needed on which identities, beyond an environmental identity, may contribute to sustainable behaviour. In this paper, we considered different identities and found positive relationships between them. This is congruent with theoretical understanding of identities as multiple (Stryker, 1980). Identities may not be disjoint and some may be complementary. We can see how an identity as health conscious may fit with a 'green' identity. Of particular interest in practice is the potential for specific behaviours to serve multiple identities: not eating meat and cycling for health reasons also serve a pro-environmental identity. This suggests that multiple identities may be of importance for environmentally friendly behaviours. This raises questions too about other identities—could an identity as 'a good citizen' or 'an upstanding member of the community' guide individuals towards more pro-environmental behaviours? Finally a fruitful path for future research will be to expand measures of identity to include factors such as attitudes, norms and self-efficacy. Such research should explore the boundaries of such factors to determine what, if any, components of attitudes, norms and self-efficacy may fall outside a conceptualisation of identity.

Acknowledgements

The research is supported by funding from the ESRC Research Group on Lifestyles Values and Environment (RESOLVE) (Grant Number RES-152-25-1004).

References

Ajzen, I. (2005) *Attitudes, personality and behavior* (2nd edn) (Berkshire, England, Open University Press).

Ajzen, I. & Fishbein, M. (1974) Factors influencing intentions and the intention-behavior relation, *Human Relations*, 27(1), 1–15.

Bamberg, S. & Möser, G. (1007) Twenty years after Hines, Hungerford, and Tomera: a new meta-analysis of psycho-social determinants of pro-environmental behaviour, *Journal of Environmental Psychology*, 27(1), 14–25.

Banerjee, B. & McKeage, K. (1994) How green is my value: exploring the relationship between environmentalism and materialism, in: C. T. Allen & D. R. John (Eds) *Advances in consumer research* (Provo, UT, Association for Consumer Research), 147–152.

Baron, R. M. & Kenny, D. A. (1986) The moderator-mediator variable distinction in social psychological research: conceptual, strategic and statistical considerations, *Journal of Personality and Social Psychology*, 51, 1173–1182.

Breakwell, G. M. (1986) *Coping with threatened identities* (London, Methuen).

Brown, K. W. & Kasser, T. (2005) Are psychological and ecological well-being compatible? The role of values, mindfulness, and lifestyle, *Social Indicators Research*, 74, 349–368.

Burke, P. J. & Reitzes, D. C. (1991) An identity theory approach to commitment, *Social Psychology Quarterly*, 54(3), 239–251.

Callero, P. L. (1985) Role-identity salience, *Social Psychology Quarterly*, 48(3), 203–215.

Clump, M. A., Brandel, J. M. & Sharpe, P. J. (2002) Differences in environmental responsibility between materialistic groups, *Psychologia*, 45, 155–161.

Conner, M. & Armitage, C. J. (1998) Extending the theory of planned behavior: a review and avenues for further research, *Journal of Applied Social Psychology*, 28(15), 1429–1464.

Corraliza, J. & Berenguer, J. (2000) Environmental values, beliefs and actions: a situational approach, *Environment and Behavior*, 32, 832–848.

Crompton, T. & Kasser, T. (2010) Human identity: a missing link in environmental campaigning, *Environment*, 52(4), 23–33.

De Groot, J. I. M. & Steg, L. (2007) Value orientations to explain beliefs related to environmental significant behavior: how to measure egoistic, altruistic and biospheric value orientations, *Environment and Behavior*, 40(3), 330.

De Groot, J. I. M. & Steg, L. (2008) Value orientations to explain beliefs related to environmental significant behavior—How to measure egoistic, altruistic, and biospheric value orientations, *Environment and Behavior*, 40(3), 330–354.

Dietz, T., Fitzgerald, A. & Shwom, R. (2005) Environmental values, *Annual Review of Environmental Resources*, 30, 335–372.

Dittmar, H. (2010) Material and consumer identities, in: S. J. Schwartz, K. Luycks & V. L. Vignoles (Eds) *Handbook of identity theory and research* (New York, Springer).

Dolnicar, S. & Grun, B. (2009) Environmentally friendly behavior, can heterogeneity be improved among individuals and contexts/environments be harvested for improved sustainable management?, *Environment and Behavior*, 41, 693–702.

Druckman, A., Chitnis, M., Sorrell, S. & Jackson, T. (2011) Missing carbon reductions? Exploring rebound and backfire effects in UK households, *Energy Policy*, 39(6), 3572–3581.

Dunlap, R. E., VanLiere, K. D., Mertig, A. G. & Jones, R. E. (2000) Measuring endorsement of the New Ecological Paradigm: a revised NEP scale, *Journal of Social Issues*, 56, 425–442.

Eagly, A. H. & Chaiken, S. (1993) *The psychology of attitudes* (Fort Worth, TX, Harcourt Brace Jovanovich College Publishers).

Evans, D. & Abrahamse, W. (2010) Beyond rhetoric: the possibilities of and for 'sustainable lifestyles', (University of Surrey).

Gatersleben, B., White, E., Jackson, T. & Uzzell, D. (2010) Values and sustainable lifestyles, *Architectural Science Review*, 53.

Hinds, J. & Sparks, P. (2008) Engaging with the natural environment: the role of affective connection and identity, *Journal of Environmental Psychology*, 28, 109–120.

Hirsh, J. B. & Dolderman, D. (2007) Personality predictors of Consumerism and Environmentalism: a preliminary study, *Personality and Individual Differences*, 43, 1583–1593.

Hitlin, S. (2003) Values as the core of personal identity: drawing links between two theories of self, *Social Psychology Quarterly*, 66(2), 118–137.

Jackson, T. (2009) *Prosperity without growth: economics for a finite planet* (Oxford, Earthscan).

Kaiser, F. G. & Wilson, M. (2004) Goal-directed conservation behavior: the specific composition of a general performance, *Personality and Individual Differences*, 36, 1531–1544.

Karp, D. G. (1996) Values and their effect on pro-environmental behavior, *Environment and Behavior*, 28, 111–133.

Kilbourne, W., Grunhagen, M. & Foley, J. (2005) A cross-cultural examination of the relationship between materialism and individual values, *Journal of Economic Psychology*, 26(5), 624–641.

Kilbourne, W. & Pickett, G. (2008) How materialism affects environmental beliefs, concern, and environmentally responsible behaviour, *Journal of Business Research*, 61, 885–893.

Lindenberg, S. & Steg, L. (2007) Normative, gain and hedonic goal frames guiding environmental behavior, *Journal of Social Issues*, 63(1), 117–137.

MacAdams, D. P. (1995) Can personality change? Levels of stability and growth in personality across the life spac, in: Weinberger Heatherton (Ed.) *Can personality change?* (Washington, DC, American Psychological Association), 299–313.

Manetti, L., Pierro, A. & Livi, S. (2002) Explaining consumer conduct: from planned to self-expressive behaviour, *Journal of Applied Social Psychology*, 32(7), 1431–1451.

McAdams, D. P. (1995) What do we know when we know a person?, *Journal of Personality*, 63(3), 365–396.

Mead, G. H. (1934) *Mind, self and society* (Chicago, IL, University of Chicago Press).

Milfont, T., Duckitt, J. & Cameron, L. D. (2006) A cross-cultural study of environmental motive concerns and their implications for pro-environmental behavior, *Environment and Behavior*, 38, 745–753.

Nigbur, D., Lyons, E. & Uzzell, D. (2010) Attitudes, norms, identity and environmental behaviour: using an expanded theory of planned behaviour to predict participation in a kerbside recycling programme, *British Journal of Social Psychology*, 49(2), 259–284.

Oreg, S. & Katz-Gerro, T. (2006) Behavior and value-belief-norm theory: predicting pro-environmental theory of planned behavior cross-nationally, *Environment and Behavior*, 38, 462–473.

Oyserman, D., Fryberg, S. A. & Yoder, N. (2007) Identity-based motivation and health, *Journal of Personality and Social Psychology*, 93(6), 1011–1027.

Pillsbury, W. B. (1934) Personality and the self, in: W. B. Pillsbury (Ed.) *The fundamentals of psychology*, (3rd edn) (New York, MacMillan Co).

Richins, M. L. (2004) The material values scale: measurement properties and development of a short form, *Journal of Consumer Research*, 31, 209–218.

Schultz, P. W. & Zelezny, L. (1999) Values as predictors of environmental attitudes: evidence for consistency across 14 countries, *Journal of Environmental Psychology*, 19(3), 255–265.

Schwartz, S. H. (1977) Normative influences on altruism, in: L. Berkowitz (Ed.) *Advances in experimental social psychology* (New York, Academic Press), 221–279.

Schwartz, S. H. (1992) Universals in the content and structure of values—theoretical advances and empirical tests in 20 countries, *Advances in Experimental Social Psychology*, 25, 1–65.

Schwartz, S. H. & Bilsky, W. (1990) Toward a theory of the universal content and structure of values: extensions and cross-cultural replications, *Journal of Personality and Social Psychology*, 58, 878–891.

Snelgar, R. (2003) Does self-identity for pro-environmental behaviours explain value-type factor structure? Paper presented at the British Psychological Society Annual Conference.

Sparks, P. & Shepherd, R. (1992) Self-identity and the theory of planned behavior: assessing the role of identification with 'Green Consumerism', *Social Psychology Quarterly*, 55(4), 388–399.

Steg, L. & Vlek, C. A. J. (2009) Encouraging pro-environmental behaviour: an integrative review and research agenda, *Journal of Environmental Psychology*, 29, 309–317.

Stern, P. C. (2000) Toward a coherent theory of environmentally significant behavior, *Journal of Social Issues*, 56(3), 407–424.

Stern, P. C. & Dietz, T. (1994) The value basis of environmental concern, *Journal of Social Issues*, 50(3), 65–84.

Stern, P. C., Dietz, R., Kalof, L. & Guagnano, G. (1995) Values, beliefs, and pro-environmental action: attitude formation toward emergent attitude objects, *Journal of Applied Social Psychology*, 25, 1161–1636.

Stets, J. E. & Burke, P. J. (2000) Identity theory and social identity theory, *Social Psychology Quarterly*, 63(3), 224–237.

Stradling, S. G., Meadows, M. L. & Beatty, S. (1999) *Factors affecting car use choices: Transport Research Institute* (Napier University).

Stryker, S. (1980) *Symbolic interactionism: a social structural version* (Menlo Park, CA, Benjamin Cummings).

Stryker, S. (1984) Identity theory—developments and extensions, *Bulletin of the British Psychological Society*, 37(SEP), A123–A123.

Thøgersen, J. (2004) A cognitive dissonance interpretation of consistencies and inconsistencies in environmentally responsible behavior, *Journal of Environmental Psychology*, 24(1), 93–103.

Thøgersen, J. & Crompton, T. (2009) Simple and painless? The limitations of spillover in environmental campaigning, *Journal of Consumer Policy*, 32, 141–163.

Thøgersen, J. & Olander, F. (2003) Spillover of environmentally-friendly consumer behaviour, *Journal of Environmental Psychology*, 23, 225–236.

Van der Werff, E., Steg, L. & Keizer, K. (2011) Values, environmental identity and pro-environmental behaviour, *Paper presented at the IAREP 2011*.

Whitmarsh, L. & O'Neill, S. (2010) Green identity, green living? The role of pro-environmental self-identity in determining consistency across diverse pro-environmental behaviours, *Journal of Environmental Psychology*, 30(3), 305–314.

Putting practice into policy: reconfiguring questions of consumption and climate change

Elizabeth Shove

Sociology, Lancaster University, Lancaster, UK

Understanding how societies change is core business for the social sciences and there is no shortage of theories about how transitions come about. Despite this reservoir of ideas, efforts to promote more sustainable patterns of consumer behaviour draw upon a remarkably narrow range of conceptual resources. The purpose of this paper is to illustrate the potential and the relevance of paradigms that lie outside the dominant discourses and traditions of economics and psychology. The method is to detail the implications of a handful of key propositions anchored in a 'strong' interpretation of practice theory. By organising this discussion around an invented conversation between a fictional policy-maker and an equally fictional social scientist, the paper explores further questions regarding the role of social theory and evidence in contemporary policy.

Introduction

It is widely agreed that the challenges of climate change are such that in the richer societies of the West many familiar ways of life and many of the patterns of consumption associated with them are fundamentally unsustainable. If there is to be any substantial and effective reduction in resource use and emissions of carbon dioxide (CO_2) new forms of living, working and playing will have to take hold. The task of understanding how social arrangements come to be as they are, and how they develop, is central to sociology, history, anthropology and material cultural studies and important for theories of socio-technical change, transition and practice. So far, few of these intellectual resources have found their way into climate change policy, much of which is dominated by efforts to nudge behaviour, modify attitudes and encourage individuals to make better, greener choices (Department for Environment, Food and Rural Affairs (DEFRA), 2008; Institute for Government, 2009).

Rather than figuring out why popular and policy debates about consumption, sustainability and everyday life rest on such a narrow slice of social science, this paper considers the potential and the limitations of just one of the many other theoretical traditions on offer. In taking this approach, it has two main aims. One is to articulate the policy implications of taking social practice rather than the actions and attitudes of individuals as the central topic of enquiry and intervention (Reckwitz, 2002; Schatzki, 2002; Warde, 2005; Shove *et al.*, 2012). The second is to illustrate some of the conceptual and practical issues that arise when moving between social theory and climate change policy. With these ambitions in mind, the paper has a rather unusual form, being organised around an imagined conversation between a climate change policy-maker (Polly) and a social scientist (Sarah). An invented dialogue between these stylised characters provides a means of introducing and structuring a discussion of how theories of practice might be mobilised, whilst providing space for further commentary on the role of social science and the meaning of evidence and relevance in the policy arena.

Before going further it is important to remember that social theories do not lead directly to prescriptions for action. In allowing us to understand the world in a particular way, they are nonetheless relevant for how policy agendas are framed and for the kinds of intervention that are deemed possible, plausible or worthwhile. In defining the problem as one of promoting pro-environmental behaviour (DEFRA, 2008), policy-related documents like *I Will If You Will* (Sustainable Consumption Round Table, 2006), *Changing Behaviour through Policy Making* (DEFRA, 2005), *Motivating Sustainable Consumption* (Jackson, 2005), and *Mindspace: Influencing Behaviour through Public Policy* (Institute for Government, 2009) reflect the prevalence of theoretical traditions from economics and psychology. In brief, there is a common understanding of behaviour as something that is driven by identifiable factors like those of rational self-interest, attitude/motivation or habit. Within this literature there are important differences of emphasis. Although many policy initiatives depend on people making rational choices, habits, frequently defined as forms of behaviour that are characterised by automaticity, frequency and a stable context, are the subject of increasing attention, in part because they complicate the impact of policies predicated on deliberate, rational action. Efforts to predict and 'nudge' (Thaler & Sunstein, 2009) habits and other forms of what Whitehead *et al.* (2011) refer to as 'more than rational' behaviour are not without their critics. Amongst others, Jones *et al.* (2010) and Whitehead *et al.* (2011) worry about the democratic legitimacy of policy interventions that are, in theory, capable of editing choices beyond the gaze of public debate and scrutiny.

As these discussions indicate, interpretations of what governments can and should do to modify behaviour vary widely, as do estimates of the relative significance of different factors including those of environment, cultural context or setting. Different approaches are nonetheless unified by the view that behaviours are outcomes of drivers, barriers and external forces, some more chosen than others (Shove, 2010). One consequence is that dominant discourses of change are situated within a bubble of intellectual space, protected and insulated from conceptual developments elsewhere in the social sciences.

Moving outside this zone implies a fundamental shift of paradigm and problem definition, but there is a growing sense that some such conceptual leap is required. The UK Committee on Climate Change (2010) has, for instance, acknowledged that since 'recent emissions reductions were far slower than those required going forward' (p. 42) a step change is required in technological innovation or behaviour change or both. The World Business Council for Sustainable Development (2009) has reached similar conclusions, also recognising that current measures are unlikely to make any really substantial difference to the carbon intensity of daily life.

Polly's puzzle

Having seen many of the reports referred to above, Polly, the fictional policy-maker, is frustrated and worried. She is well aware of the scale of CO_2 emissions and of the rate at which these need to drop if there is to be any chance of meeting current targets. She has had a hand in developing persuasive strategies and financial measures to change consumer behaviour and knows that these have had limited impact. She puts her concerns to Sarah, a social scientist of her acquaintance:

> Polly: The last decade of concentrated effort on behaviour change has not changed anything very much at all. My colleagues and I have tried driving public behaviours towards more sustainable lifestyles but to no effect. What is going wrong, what can I do next?

Sarah responds but not in the way that Polly expects. This is what she says:

> Sarah: Perhaps you should reframe your problem: what if you forgot about persuading individuals to use less energy and water and concentrated on how resource intensive practices take hold in society and on how they change? Surely that is the key question.

Warde (2005) suggests that consumption is usefully understood as an outcome of the routine reproduction of ordinary practices. Sarah's advice clearly builds on this idea. But what does Sarah mean by 'practice'? Before taking the conversation further more should be learnt about her interpretation of this concept.

Sarah's concept of social practice

In the last few years, authors of articles on energy demand and sustainable consumption and have begun to write about practices. For example, Gram-Hanssen (2010) writes about the practice of standby consumption (meaning the practice of keeping appliances on standby); Crosbie & Guy (2008) consider lighting practices; Wilhite (2008) refers to energy practices, and Strengers (2011) to those that demand water. In most of these cases the terminology of practice signals affiliation to a loose tradition of sociotechnical approaches and distance from behavioural accounts emphasising attitudes and values. These authors make much of the fact that consumer choices are constrained and 'scripted' (Akrich, 1992) by material context and environment. In essence they contend that practices—what individuals do—reflect the pursuit of shared goals (comfort, mobility) within a particular sociotechnical setting. As represented by Gram-Hanssen (2010), the value of practice theory is that 'it empha-sizes sociotechnical structures as the basis for analyzing stability of consumer practices and opportunities for change' (p. 150). From this perspective, invoking theories of practice is more or less the same as invoking concepts of sociotechnical change in which practice/user behaviour is shaped by, and co-evolves with, relevant aspects of infrastructure, culture and design.

Others writers refer to practice as a means of developing a fuller, more comprehen-sive account of individual behaviour. This is the approach adopted by Hargreaves (2011) who suggests that practice theory is of value in that it 'provides a more holistic and grounded perspective on behaviour change processes as they occur in situ' (p. 79). In this case, reference to practice is liberating in that it legitimises reference to an unu-sually wide range of driving factors, including social relations, social norms and insti-tutional contexts. 'In so doing, it offers up a wide range of mundane footholds for behavioural change, over and above individuals' attitudes or values' (p. 79). In Hargreaves's article, practice theory provides a convenient label with which to badge approaches that take social action to be constructed, situated and performed. For Halkier & Jensen (2011), this latter feature is crucial. They conclude that the dis-tinctiveness of 'practice theory is that the performative character of social life is fore grounded and privileged analytically' (p. 103). Meanwhile, Barr *et al.* (2011) apply practice approaches, again very loosely defined, to the task of understanding localised instances of behaviour, as captured through case studies and interviews. In much of this writing the central project remains that of understanding how and why people act as they do.

By contrast, Sarah takes an altogether stronger line and takes *practices* as the central topic of her enquiry (Giddens, 1984, p. 3). This might sound like a subtle distinction but it is one that matters for the types of questions she asks, and for how she goes about her work. In keeping with this orientation, Sarah is interested in understanding how practices emerge, persist and disappear. By practices she means recognisable entities that exist across time and space, that depend on inherently provisional inte-grations of elements, and that are enacted by cohorts of more and less consistent or faithful carriers.

Unlike those who talk of practices as a means of talking about materialised, situated moments of performance, Sarah has another agenda. She is primarily interested in the development of practices-as-entities, and in their distribution across space and time (Giddens, 1984). By practices-as-entity she takes a practice (for example, playing football, daily showering, commuting, etc.) to be something that exists between and beyond specific moments of enactment. Defined like this, practices are carried, sustained and transformed by cohorts of practitioners (those who do). Practices-as-entities would not exist without reproduction, and reproduction depends on localised instances of performance. Even so, it is both possible and, in Sarah's view, useful to focus not on the people who do the enacting, but on the practice that they repro-duce/transform. Sarah is consequently interested in how certain practices manage to secure carriers or hosts who are willing and able to devote significant resources of time and energy to reproducing them over and over again. Rather than trying to

understand habit as a form of behaviour that people adopt, she is, for instance, interested in how habits capture and retain cohorts of suitably devoted practitioners
(Shove, 2011).

While many turn to a vocabulary of 'practice' (as opposed to behaviour) as a means
of signalling the socially constructed nature of action, Sarah believes that concepts
developed by Giddens (1984), Schatzki (2002) or Reckwitz (2002) have further,
deeper implications for the analysis of change. She is convinced that these theoretical
resources can be used to describe inherently dynamic processes in which the constitutive elements of practice (the meanings, competences, materials) integrated in
each performance are themselves subject to change (Shove & Pantzar, 2005), and
in which the margins of practices-as-entities extend and shrink as new carriers are captured, and as others defect. In short, her interest in the changing contours of practices-
as-entities sets her approach apart from those who invoke theories of practice as a
means of enriching knowledge of consumer behaviour.

To come back to questions of climate change, the key issue for Sarah is to understand
the trajectories and careers of variously resource intensive practices (as entities). From
her point of view this is a matter of identifying the elements of which such practices are
made, learning about their history (since elements are themselves outcomes of practices
past) and about also crucial processes of recruitment and defection: how are people
drawn into more or less sustainable practices and how do their lives and careers
sustain the lives and careers of the practices they reproduce? In theory, the policy implications of such an approach are relatively clear: engendering long-term transformation
in what counts as a normal and acceptable way of life depends on reconfiguring the
elements of practice; relations between practices, and patterns of recruitment and
defection. Let us now resume the conversation, allowing Sarah to continue.

Reconfiguring the elements of practice

Sarah: As I was about to explain, social practices—for example, driving or cycling to work
or cooking and eating dinner—involve the active integration of 'elements'. These include:
materials, objects and infrastructures; forms of competence and know-how, images and
meanings.

Polly: Steady on! I am interested in changing people's behaviour and encouraging them to
adopt a more sustainable way of life but you are talking about elements and practices, I
don't get it. I need practical advice. Just tell me, what should I do?

Sarah: Your job, as a policy maker, is to influence both the elements of existing practices—to make them more sustainable—and to think about the total range of practices
that might make up a more sustainable society.

Polly pauses for a moment before responding:

Polly: So far, my job has been that of persuading individuals to consume less. You are now
saying that if I want to reduce water and energy consumption I ought to think about the
materials, meanings and competences of which practices like daily showering are made,
and how they change?

In this snippet of conversation, Sarah turns the problem away from that of individual
behaviour towards an understanding showering as an emergent, historically specific,

outcome of the interweaving of running hot water, bathrooms, concepts of freshness and invigoration, and taken-for-granted skills of personal care (Hand *et al.*, 2005). It is the repeated integration of these elements that makes showering such a regular and normal pursuit for so many people today. Intrigued by this curious perspective Polly begins to wonder. If the constitutive elements of unsustainable practices were not in circulation, what would become of the practices of which they are part? Could governments edit 'bad' elements, and hence 'bad' practices out of existence? This sounds far too radical and far too much like top-down state intervention, but at the same time Sarah has a point. Polly and her colleagues, past and present, clearly have a hand in structuring the materials, meanings and competences around which daily lives revolve.

In areas like public health there is a long and respectable tradition of combining investment in infrastructures (sewerage systems, mains water) with campaigns instilling techniques such as those of washing regularly, along with ideas about what it is to be clean (Ogle, 1996; Melosi, 2000). Similarly, post-war urban planning was, in many ways, about furnishing the ingredients of which desired ways of living might be made. Tapiola, a Finnish town designed and built in the early 1960s, was inspired by Ebenezer Howard's vision of the model garden city (Howard *et al.*, 1951); by Patrick Geddes' belief that spatial form could be used for social ends (Geddes, 1915), and by Lewis Mumford's ideas of technological progress (Mumford, 2010). In setting out streets, shops, homes and schools, Tapiola's planners were clear about what they wanted to achieve. Their aim was to create conditions in which children ran free, the community was strong, the interaction with nature was easy, and family life was close, harmonious and healthy. Ideological visions of the good life were inscribed and materialised in the smallest detail of kitchen design through to the distance between home and school (Hertzen & Spreiregen, 1971). Not everything went to plan, but there is no doubt that the ambition was to bring new social arrangements into being by providing the moral and material infrastructure around which they might develop.

The idea that daily lives might be scripted on such a scale, and with such precision, has fallen out of fashion and it is in any case clear that designers and nation-states have limited ability to control the circulation and flow of ideas about what it is to be modern or what a successful life entails. Many elements of practice circulate in ways that show scant regard for national borders. In addition, and in areas like food consumption or building design, global systems of provision are important in structuring diets and meals and in configuring the architecture of urban living. National policy-makers like Polly can do only so much to promote or stem the transnational diffusion of materials, meanings and forms of competence, but as Sarah's next example illustrates, there may be ways of intervening in how practices are constituted and in the forms of energy and resource consumption they require:

> Sarah: Changing elements of practice is not just a matter of engineering and planning. Take a look at what has been happening in Japan—in 2005 the government introduced a programme called 'Cool Biz', modifying conventions and practices of clothing as a means of reducing energy demand in the hotter months of the year.

Polly wants to know more, so Sarah provides a brief account.

Air-conditioning technologies have made it possible to manipulate humidity, temperature and ventilation and have been crucial in defining and diffusing standardised concepts of comfort and conventions of normal and appropriate clothing. In fewer than 70 years, methods of defining and calculating optimal indoor conditions, initially developed in Northern Europe and the United States, have been appropriated and copied around the globe: it has become normal to heat or cool buildings to a steady 22°C whatever the weather outside. Vast quantities of energy are already consumed in maintaining these conditions and there is scope for more. According to Sivak (2009: 1382), 'the potential cooling demand in metropolitan Mumbai is about 24% of the demand for the entire United States' (Isaac & van Vuuren, 2009). These are scary figures and the question faced by Polly and her colleagues around the world is whether they can break this vicious circle of energy demand by redefining the *elements* of comfort.

In 2005 the Japanese government took a step in this direction. The idea was simple: government buildings would not be heated or cooled between 20 and 28°C, and male office workers would be encouraged to remove jackets and ties in the summer and wear more in the winter (called 'Warm Biz'). The effect was to change the *meaning* of normal clothing, along with the technologies (levels of air-conditioning) and competences (of dress and of facilities management) involved in the routine enactment and effective accomplishment of office life. By most measures 'Cool Biz', the summer variant, has been spectacularly effective, resulting in an estimated 1.4 million tonnes-reduction in CO_2 emissions (Knee Tan *et al.*, 2008) and making a tangible difference to what men and women wear at work and to a lesser extent in the home.

This strategy appears to have transformed collective conventions rapidly and on a significant scale. The Cool Biz programme worked on a number of fronts at once. Established marketing techniques were used to transform the *meaning* of smart and appropriate wear. The then prime minister, Mr Junichiro Koizumi, and members of the Cabinet were shown wearing loose-fitting short-sleeved outfits in formal settings. Successful business leaders were involved, the clothing industry responded to the challenge and large department stores promoted especially designed garments under the Cool Biz name.

Although not inspired by social theories of practice and not positioned as a rejection of unsustainable conventions imported from the West, Cool Biz appears to have transformed expectations of indoor climates and of what it is acceptable to wear, and to have changed both in a direction that reduces energy consumption and CO_2 emissions (De Dear, 2007).

Polly and Sarah are both impressed by Cool Biz, but for different reasons. Polly takes it to be a surprisingly successful example of remarkably effective social marketing. For Sarah, the fact that Cool Biz resulted in a transition in practice was more by accident than intent. Even so, it demonstrates that policy-makers can intervene to shape the elements of office life and can do so on a societal scale. It also shows that programmes designed in terms of one paradigm (individual persuasion) can have unintended consequences, inadvertently reconfiguring the elements of practice.

To date, Cool Biz remains at the level of fashion and has yet to be embedded in building codes or estimates of future energy demand. From Sarah's point of view, it consequently falls short of its potential as an intervention in practice. Polly has other concerns:

> Polly: As far as I can see, Cool biz is an interesting case, but office workers taking off ties is not going to change the world. My job is to reduce CO_2 emissions on a massive scale, and to do so fast. How can you help with that? How can I persuade people to leave their cars at home? Or to eat less meat?

Sarah struggles to disguise her frustration:

> Sarah: I don't think you will get very far if you continue to define your job as one of persuading individuals to change their ways, one by one.

Polly has another go:

> Polly: Alright, let me try another question. How can I use your ideas to foster transitions not in just in one practice, but in many at once?
> Sarah: OK, that is a critical issue and yes, there is certainly more to say about how practices relate to each other. Let's talk about cars and bikes.

Before commenting on how Polly might act to forge or break links *between* more and less sustainable practices, Sarah begins by describing the changing relation between systems of velo and automobility.

Reconfiguring relations between practices

Cycling is now widely recognised as a practice that is good for the environment and for personal health: Sustrans (2008) claims that 2 kg of carbon are saved for every short journey made by bicycle. Cycling is also a means of transport that used to be very much more widespread than it is today. In 1949 in the UK, an estimated '34 per cent of all mechanised journeys were made by bicycle. Fifty years later that figure had fallen to 2 per cent' (*The Times*, 2008). By any standards this is a spectacular decrease, representing a rapid and radical movement *away* from what used to be a normal and familiar (low carbon) practice, namely that of riding a bicycle to work.

Although this decline coincided with the rise of the car as an increasingly democratic means of personal mobility, the narrative is not one of simple substitution. As Geels (2005) explains, the development of cycling laid the foundations for many of the elements on which the coming system of automobility depended, including aspects of infrastructure (road surfaces, production capacity) along with ideas and expectations of personal mobility.

Histories of driving and cycling show that the relation between them is inherently dynamic, and important for the trajectories of both. Whether cycling is characterised as slow, dangerous or demanding is not just a matter of personal opinion, but is instead related to the systemic configuration of this practice and of others in terms of which it is defined. For example, in the 1940s, and when compared with walking, cycling provided a *fast* means of covering extended distances. These qualities are relative and when cycling takes place in urban environments designed around cars, or when daily routines involve travelling distances only made possible by the car, cycling is redefined as slow, effortful and inconvenient. In short, interpretations of cycling as a normal or an unusual thing to do depends on how riding is positioned within and by an interdependent network of social and material arrangements.

These ideas are useful in making sense both of the rapid decline of cycling in many European countries between the 1950s and 1970s and of its resurgence in some locations but not in others. According to Pucher and Buehler 'the bike share of trips fell from 50%–85% of trips in 1950 to only 14–35% of trips in 1975 in a range of Dutch, Danish and German cities' (2008, p. 502). In some European countries, rates of cycling have increased sometimes by as much as 20% since the mid-1970s, but in others, like the UK, the modal share has remained more or less unchanged at around 1% for the last 40 years (Cabinet Office Strategy Unit, 2009).

This variation is intriguing: what Urry (2004) refers to as the 'system of automobility' is no less established in Denmark or the Netherlands than in the UK. In these as in other countries, the petrol and steel car has been systematically locked in to the organisation of society. Cars have become progressively embedded through patterns of economic and suburban development and through spatial and temporal arrangements that demand and assume a relentless logic of automobility. However, generic trends in automobility disguise important local variation in the extent and degree to which alternative regimes (including those of cycling) coexist (de la Bruheze, 2000).

In the Dutch case the persistence of relevant meanings, competences and bicycle-related infrastructures seems to have made it easier to reinstate cycling at least to some degree. Reflecting on why similar efforts have met with limited success in the UK, de la Bruheze (2000) suspects that it might be 'because bicycle use had declined too far' and because the 'material and social bicycle culture had disappeared' (p. 4). These examples suggest that in some circumstances the elements of cycling-as-normal endure, but in dormant form, and that in others it is not just that requisite links are temporarily broken but that vital ingredients have actually disappeared. In such settings:

> Attempting to reform technology without systematically taking into account the shaping context and intricacies of internal dynamics may well be futile. If only the technical components of a system are changed, they may well snap back into their earlier shape like charged particles in a strong electromagnetic field. The field also must be attended to; values may need to be changed, institutions reformed, or legislation recast.(Hughes, 1983/1993, p. 465)

For policy-makers seeking to engender change, the two situations described above— one in which links are broken but where relevant elements still exist (as in Denmark and the Netherlands), the other (the UK) in which necessary elements are not (or no longer) in place—call for quite different forms of intervention.

This far, Sarah's reflections on driving and cycling draw attention to the changing relation between practices, to the potential for symbiotic as well as competitive relationships, and to the consequences of past configurations for the future. These thoughts prompt Sarah to remind Polly that currently dominant systems of automobility exist alongside the remains of past, partly dormant or largely extinct complexes of practice and that policy interventions take place within and not outside these locally specific histories.

She makes this point by comparing policy interventions to promote more sustainable transport in London and Groningen in the Netherlands. In Groningen, almost 40% of local trips are currently made by bicycle, a situation in part shaped by long-term political commitment to cycling through mutually reinforcing policies of compact land-use planning, schemes to restrict car use and investment in cycling infrastructure. In this city, programmes designed to reduce transport-related carbon emissions have an effect in an environment in which cycling is already normal and mundane.

By contrast, efforts to promote cycling in London take effect in a situation in which riding a bicycle is *not* an entirely familiar thing to do. The congestion charge (introduced in 2003—car drivers pay to enter the Central London charging zone during certain hours of the day) and direct investment in bicycle routes have coincided with rapid and recent recruitment. Rates of cycling in London increased by at least 50% between 2003 and 2007 and continue to grow. Although this is from a very low base of 1–2% of modal share, the current pace of change might imply that certain positive feedback effects are underway: that cycling is quickly becoming more normal as the practice captures more normal people.

The prospect of deliberately engineering the demise of automobility and the rise of cycling as a newly dominant form might be a distant one, but in thinking about the potential for reconfiguring relations between practices, Sarah identifies two key points. First, in so far as they make a difference, policy initiatives do so not in the abstract, but to processes that already have a life and a history of their own. Political opportunities for intervention, and the form these take, are emergent effects of the very systems that policy-makers seek to influence. Second, where such interventions reconfigure the relation between practices, for example systematically prioritising bicycles over cars, they can set in train processes of positive feedback the effects of which are unpredictable in terms of extent (for example, regarding the scale of recruitment) and depth (for example, how firmly new configurations become embedded).

> Polly, who has been listening quietly, has another question for Sarah:
> Polly: I get what you are saying, but where do people come into the story?
> Sarah: People are vital as the carriers and transformers of practice. If practices like driving or cycling are to survive they have to secure and maintain resources and practitioners willing and able to keep them alive.
> Polly: That is all very well but what can I do to increase the chances that people will be captured by sustainable practices?

Reconfiguring paths and projects

Now it is Sarah's turn to pause. There is a lot to think about here and there are various threads to follow. It is clear that she needs to talk about how forms of access and participation are structured by policy, not just now, but also by the cumulative effects of policies enacted in the past. But where should she begin?

At the most basic level, the probability of encountering and participating in different practices is structured by divisions like those of age, gender and social class. The suggestion that governments should enhance what Dahrendorf (1979) refers to as 'life chances' (meaning an individual's opportunity to maximise his or her own talents) acknowledges inequities of access and distribution, some of which are rooted in patterns of advantage built up over many generations. It also raises further questions about the range of practices to which people aspire, about what talents count and what it means to maximise them. For Bourdieu (1984), the idea of *habitus* provides a means of bridging between the cumulative (and unequal) effects of past experiences, resources, dispositions and tastes, and the content and character of *future-oriented* aspirations and opportunities. This is a theme also explored by Pred (1981) who writes about how

> the particular economic and cultural practices in which individuals of a given group or class partake appear 'natural', 'sensible', or 'reasonable', even though there is no awareness of the manner in which those practices are either adjusted to other practices, or structurally limited. (p. 8)

Pred's conclusion that definitions of valued pursuits are themselves outcomes of dialectical interaction between individual and institutional projects is crucial. It is so in that it supports the conclusion that 'social transformation and altered structural relations can only occur through the introduction, disappearance or modification of institutional projects' (p. 17). The point is that individuals' daily and life paths are intertwined with collective, institutional projects to which they lend time and energy, for instance through roles such as those of employee, parent, etc. Participation is, in turn, relevant for the direction that individual life paths take, and for the kinds of experience and expertise acquired along the way. Past performances are evidently vital for the accumulation of know-how and competence and for the emergence of institutional projects, some of which are rather more sustainable than others.

Sarah knows all this but she is still not sure how to respond. What do these observations mean for policy-makers and others seeking to promote more sustainable practices?

At the broadest, most 'macro'-level, dominant institutional projects (i.e. those which command time, resources and attention) are complex amalgams of past trajectories and current aims and aspirations, many of which are materially sustained and reinforced by the state. Issues of sustainability appear in many guises when approached at this scale, and when considered with reference to the reproduction of social institutions, including conventions of family life, systems of provision and consumption, economic relations and more. In so far as governments have a hand in reproducing these institutions and systems, and the versions of normal and acceptable ways of life associated with them, they also have a hand in configuring related patterns of mobility and resource consumption.

To give one very specific example, the idea that parents should have a choice of schools has generated more moving around than was the case when children simply went to the school that was closest to home. Other self-fulfilling conventions of need and entitlement are tacitly buried in plans and strategies for energy supply and in the design of resilient water infrastructures. Practices and associated 'standards of living' are, in effect, inscribed in how infrastructures are conceptualised and managed. Through arrangements like these variously unsustainable institutional projects are tacitly reproduced all the time, not at the forefront of explicit policy intervention but as part of the backdrop of taken-for-granted order: this being an order structured around specific bundles and complexes of practice.

It is in these terms that Sarah replies:

> Sarah: You could think about how different areas of public policy (education, health, family, work, leisure etc.) inadvertently increase the chances that people will be captured by unsustainable practices. And you could reflect on more fundamental questions about the sustainability, or otherwise, of normal policy goals.
>
> Polly: But that is a massive task, and one that is well beyond my reach. What you are talking about is a systemic review of the unintended consequences of just about every area of government policy. And you are also talking quite directly about questions of politics and power.

In this exchange Sarah and Polly stumble over a number of limiting conditions. From Sarah's point of view, she has done her best to introduce a handful of ideas rooted in a strong interpretation of practice theory and has tried to demonstrate their relevance for conceptualising and promoting new ways of living on the scale required if there is to be any really significant reduction in CO_2 emissions. However, Polly, who simply wanted advice about how to help people make 'better', more sustainable choices, finds herself drawn in to a rather challenging discussion of how policy-making structures patterns of consumption.

Reconfiguring the conversation

It was not what she bargained for, but Polly is actually quite excited by Sarah's account of how state actors influence the distribution and circulation of materials, competences and meanings, and how governments have a hand in forging and breaking some of the links involved in the surprisingly uncontrollable, surprisingly living system that is daily life. In catching sight of these dynamic processes, Polly caught sight of a new future: rather than persuading individuals to change their behaviour, one person at a time, she could be out there building networks and coalitions and constructing partnerships that make the conditions of sustainable practice possible. For Polly, the novelty is not in recognising that infrastructures and social networks matter. This is not in itself news: having flicked through many government reports, she is aware of attempts to marry notions of cultural capital, social networks and environmental circumstance with behavioural policy (Knott *et al.*, 2008). Rather, the key insight, and the one that has really caught Polly's imagination, has to do with how change is conceptualised and with how she thinks about her own role.

Instead of looking for the drivers and barriers of individual behaviour, whether in cultural context, infrastructure or elsewhere, Polly is becoming interested in the dynamics of social practice, as such. Can she shape trajectories and promote and hinder the development of more and less sustainable practices-as-entities? There is a limit to what she can do, and in any case, practices have lives of their own. Because of this, the practical consequences of Polly's interventions are likely to be unstable and unpredictable in that the practices they seek to shape are subject to ongoing reproduction/transformation. But from what Sarah has said, this does not rule out the possibility of thoughtful, practice-oriented strategies: if she plays her cards right, Polly might be able to increase the chances that lower carbon ways of life persist and thrive.

If she is to take these ideas to heart, Polly will have to redefine the agendas and priorities of those with whom she works. At the moment she spends a lot of time and money surveying individual responses to batteries of attitudinal questions about the environment. But is this really the sort of information she needs if the aim is to understand and potentially shape the range of practices of which contemporary society is formed? Probably not. If she redefines the problem, other sorts of data, and other styles of enquiry, will be required. These might include concerted and innovative efforts to quantify the growth of certain practices and the demise or transformation of others. Radical reduction in CO_2 emissions implies that conventions, standards, routines, forms of know-how, markets and expectations will need to change on a massive scale. Could Polly develop methods of detecting and quantifying systemic moves in this direction? Could she come up with some cross-sectoral analysis of how policy-making of all forms influences the texture and rhythm of daily life and the patterns of consumption on which such arrangements depend?

Wilson & Chatterton (2011) argue that policy-makers pick and mix from menus of conceptually incommensurable strategies and approaches, selecting a tool from here and a measure from there in pursuit of methods that 'work'. Polly is not committed to conceptual consistency for its own sake, but she is decidedly uneasy about flitting between practice-oriented and behavioural models, as if these provide different takes on *the same* phenomenon. It is true that practice-oriented policy might draw on similar methods and techniques (the scope of what government can do remains constrained), but for Polly, the crux of the matter, and the excitement, is that a practice orientation is strategically important. In very practical terms, the priorities that matter when the aim is that of promoting pro-environmental behaviour are not the same as those that pertain when the goal is one of reconfiguring the practices that people reproduce.

If Polly is to figure out how the state sustains unsustainable institutions, conventions and ways of life, and if she is to exploit opportunities for fostering other options and possibilities she will have to extend the range of social theory on which she draws. For the moment, policy relevant social science is that which is consistent with a dominant paradigm organised around theories of individual attitude, behaviour and choice, but if this tide should turn there would be no harm in giving Sarah a call and having a rather longer conversation about what more the social sciences have to offer.

Acknowledgements

This article draws on an 'Extraordinary Lecture' presented and performed by members of the 'Social change-climate change' working party and on research funded as part of Elizabeth Shove's ESRC Climate Change Leadership Fellowship, funded by the ESRC. Award No: RES 066 27 0015. It is also based on the final chapter of Shove, E, Pantzar, M. and Watson, M. (2012) *The Dynamics of Social Practice: Everyday life and how it changes*, published by Sage. Permission to use some of this material is gratefully acknowledged.

References

Akrich, M. (1992) The de-scription of technical objects, in: W. Bijker & J. Law (Eds) *Shaping technology/building society* (Cambridge, MA, MIT Press), 205–224.

Barr, S., Shaw, G. *et al.* (2011) Sustainable lifestyles: sites, practices, and policy, *Environment and Planning A*, 43(12), 3011–3029.

Bourdieu, P. (1984) *Distinction: a social critique of judgement and taste*. London: Routledge.

Cabinet Office Strategy Unit (2009) *An analysis of urban transport*. London: HMSO.

Committee on Climate Change (2010) *Meeting carbon budgets—ensuring a low-carbon recovery: second progress report*. Available online at: http://downloads.theccc.org.uk.s3.amazonaws.com/0610/CCC-Progress-Report-web-version_3.pdf (accessed on 3 January 2012).

Crosbie, T. & Guy, S. (2008) Enlightening energy use: the co-evolution of household lighting practices, *International Journal of Environmental Technology and Management*, 9, 220–235.

Dahrendorf, R. (1979) *Life chances: approaches to social and political theory*. Chicago, Chicago University Press.

De Dear, R. J. (2007) Comments on 'Clothing as a mobile environment for human beings—prospects of clothing for the future' presented by Teruko Tamura, Presidential Address to ICHES'05 Tokyo, Japan 12–15 September 2005, *Journal of the Human–Environmental System*, 10(1), 45–46.

De la Bruheze, A. (2000) *Bicycle use in twentieth century Western Europe: the comparison of nine cities*. Available online at: www.velomondial.net/velomondiall2000/PDF/BRUHEZE.PDF (accessed 4 June 2012).

Department for Environment, Food and Rural Affairs (DEFRA) (2005) *Changing behaviour through policy making* (London, HMSO).

Department for Environment, Food and Rural Affairs (DEFRA) (2008) *A framework for pro-environmental behaviours* (London, HMSO).

Geels, F. W. (2005) *Technological transitions and system innovations: a co-evolutionary and socio-technical analysis* (Cheltenham, Edward Elgar).

Geddes, P. (1915) *Cities in Evolution* (London, Williams and Norgate).

Giddens, A. (1984) *The constitution of society* (Cambridge, Polity).

Gram-Hanssen, K. (2010) Standby consumption in households analysed with a practice theory approach, *Journal of Industrial Ecology*, 14(1), 150–165.

Halkier, B. & Jensen, I. (2011) Methodological challenges in using practice theory in consumption research. Examples from a study on handling nutritional contestations of food consumption, *Journal of Consumer Culture*, 11(1), 101–123.

Hand, M., Shove, E. & Southerton, D. (2005) Explaining showering: a discussion of the material, conventional, and temporal dimensions of practice, *Sociological Research Online*, 10(2).

Hargreaves, T. (2011) Practice-ing behaviour change: applying social practice theory to pro-environmental behaviour change, *Journal of Consumer Culture*, 11(1), 79–99.

Hertzen, H. v. & Spreiregen, P. D. (1971) *Building a new town: Finland's new garden city, Tapiola* (Cambridge, MA, MIT Press).

Howard, E. S. (1951) *Garden cities of tomorrow* (London, Faber).

Hughes, T. (1993 [1983]) *Networks of power: electrification in Western society, 1880–1930*. Baltimore, MD, Johns Hopkins University.

Institute for Government (2009) *Mindspace: influencing behaviour through public policy*. Available online at: http://www.instituteforgovernment.org.uk/images/files/MINDSPACE-full.pdf.

Isaac, M. & van Vuuren, D. P. (2009) Modeling global residential sector energy demand for heating and air conditioning in the context of climate change, *Energy Policy*, 37(2), 507–521.

Jackson, T. (2005) *Motivating sustainable consumption*. Available online at: http://www.sd-research.org.uk/post. php?p=126 (accessed on 3 January 2012).

Jones, R., Pykett, J. *et al.* (2011) Governing temptation: changing behaviour in an age of libertarian paternalism, *Progress in Human Geography*, 35(4), 483–501.

Knee Tan, C., Ogawa, A. & Matsumura, T. (2008) *Innovative climate change communication: team minus 6%*, GEIC Working Paper Series No. 2008-001. (Tokyo: United Nations University).

Knott, D., Muers, S. & Aldridge, S. (2008) *Achieving culture change: a policy framework strategy unit* (London, HMSO).

Melosi, M. (2000) *The sanitary city: urban infrastructure in America from colonial times to the present* (Baltimore, MD, Johns Hopkins University Press).

Mumford, L. (2010 [1963]) *Technics and civilization* (Chicago, Chicago University Press).

Ogle, M. (1996) *All the modern conveniences: American household plumbing, 1840–1890* (Baltimore, MD, Johns Hopkins University Press).

Pred, A. (1981) Social reproduction and the time-geography of everyday life, *Geografiska Annaler. Series B. Human Geography*, 63(1), 5–22.

Pucher, J. & Buehler, R. (2008) Making cycling irresistible: lessons from the netherlands, denmark and germany, *Transport Reviews*, 28(4), 495–528.

Reckwitz, A. (2002) Toward a theory of social practices: a development in culturalist theorizing, *European Journal of Social Theory*, 5(2), 243–263.

Schatzki, T. R. (2002) *The site of the social: a philosophical account of the constitution of social life and change* (University Park, PA, Pennsylvania State University Press).

Shove, E. (2010) Beyond the ABC: climate change policy and theories of social change, *Environment and Planning A*, 42(6), 1273–1285.

Shove, E. (2011) Habits and their creatures, paper presented at *Social Science and Sustainable Consumption*, Helsinki, Finland, 27–29 April 2011.

Shove, E. & Pantzar, M. (2005) Consumers, producers and practices: understanding the invention and reinvention of Nordic Walking, *Journal of Consumer Culture*, 5(1), 43–64.

Shove, E., Pantzar, M. & Watson, M. (2012) *The dynamics of social practice: everyday life and how it changes* (London, Sage).

Sivak, M. (2009) Potential energy demand for cooling in the 50 largest metropolitan areas of the world: Implications for developing countries, *Energy Policy*, 37(4), 1382–1384.

Strengers, Y. (2011) Beyond demand management: co-managing energy and water practices with Australian households, *Policy Studies*, 32, 35–58.

Sustainable Consumption Round Table (2006) *I will if you will*. Available online at: http://www.sd-commission.org.uk/publications.php?id=367 (accessed 4 June 2012).

Sustrans (2008) *Why cycle?* Available online at: http://www.sustrans.org.uk/assets/files/leaflets/sustrans_whycycle_March08.pdf (accessed on 3 January 2012).

Thaler, R. H. & Sunstein, C. R. (2009) *Nudge: improving decisions about health, wealth and happiness* (London, Penguin).

The Times (2008) Ian Hibell, cyclist who pedalled world, killed by hit-and-run driver, *Times Online*. Available online at: http://www.thetimes.co.uk/tto/public/sitesearch.do?querystring=Ian+Hibell%2C+cyclist+who+pedalled+world%2C+killed+by+hit-and-run+driver&p=tto&pf=all&bl=on (accessed 4 June 2012).

Urry, J. (2004) The 'System' of Automobility, *Theory, Culture & Society*, 21(4–5), 25–39.

Warde, A. (2005) Consumption and theories of practice, *Journal of Consumer Culture*, 5(2), 131–153.

Whitehead, M., Jones, R. & Pykett, J. (2011) Governing irrationality, or a more than rational government? Reflections on the rescientisation of decision making in British public policy, *Environment and Planning A*, 43(12), 2819–2837.

Wilhite, H. (2008) New thinking on the agentive relationship between end-use technologies and energy using practices, *Journal of Energy Efficiency*, 1(1), 121–130.

Wilson, C. & Chatterton, T. (2011) Multiple models to inform climate change policy: a pragmatic response to the 'beyond the ABC' debate, *Environment and Planning A*, 43(12), 2781–2787.

World Business Council for Sustainable Development (WBCSD) (2009) *Energy efficiency in buildings*. Available online at: http://www.wbcsd.org/Pages/EDocument/EDocumentDetails.aspx?ID=13559&NoSearchContextKey=true (accessed on 3 January 2012).

Input–output analyses of the pollution content of intra- and inter-national trade flows

Karen Turner[a], Cathy Xin Cui[b], Soo Jung Ha[c] and Geoffrey Hewings[d]

[a]Division of Economics, Stirling Management School, University of Stirling, Stirling, UK; [b]Fraser of Allander Institute, Department of Economics, University of Strathclyde, Glasgow, UK; [c]Korean Research Institute for Human Settlements, Soeul, South Korea; [d]Regional Economics Applications Laboratory, University of Illinois, Urbana-Champaign, IL, USA

This paper considers the application of input–output accounting methods to consider the pollution implications of different production and consumption activities, with specific focus on pollution embodied in intra- and inter-national trade flows. It considers the illustrative case studies of production and consumption measures of emissions and pollution embodied in interregional trade flows between two regions of the UK and between five Midwest regions/states within the United States. The analysis raises questions in terms of policy reliance on the extremes of conventional production and consumption accounting measures and considers a range of alternative measures that may be calculated using input–output methods to provide different informational content. The paper focuses on different types of air pollutant of current policy concern in each the UK and the US Midwest cases and demonstrates how use of the environmental input–output framework allows the analysis of the nature and significance of interregional pollution spillovers. The results raise questions in terms of the extent to which authorities at the regional level can limit local emissions where they are limited in the way some emissions can be controlled, particularly with respect to changes in demand elsewhere within the national economy. This implies a need for policy coordination between national and regional level authorities to meet emissions reductions targets. Moreover, the existence of pollution trade balances between regions also raises issues regarding net losses/gains in terms of pollutants as a result of interregional trade. In conducting analyses for different types of air pollutant (here carbon dioxide, CO_2, as a global warming gas, a greenhouse gas (GHG), in the UK case; and ammonia, NH_3, as a pollutant of more local concern in the US case) the paper also considers how pollution embodied in international trade flows may be accounted for and attributed.

Introduction to production and consumption emissions accounting in the literature

A crucial issue impacting on unilateral attempts to fulfil national emissions reductions targets under the Kyoto Protocol is the impact of trade on any individual region's or country's domestic emissions generation. One problem is that the generation of emissions in producing goods and services to meet export demand is charged to the producing nation's emissions account. Munksgaard & Pedersen (2001) highlight this issue, distinguishing between a 'production accounting principle' (PAP) and a 'consumption accounting principle' (CAP). The former focuses on emissions produced within the geographical boundaries of the national economy. This is what is accounted for, and what individual national governments are responsible for reducing, under the Kyoto Protocol. In contrast, the latter focuses on emissions produced globally to meet consumption demand within the national economy. This is what increasingly popular measures such as carbon footprints attempt to measure, and what many people regard as more appropriate, given that human consumption decisions are commonly considered to lie at the heart of the climate change problem.

An extensive discussion on the allocation of greenhouse gas (GHG) emissions has been conducted in the literature (e.g. Wyckoff & Roop, 1994; Kondo *et al.*, 1998; Ferng, 2003; Bastianoni *et al.*, 2004; Mongelli *et al.*, 2006). In parallel to this discussion there has been a development of models and accounting techniques that can account for pollution embodied in trade, and this has mainly involved the use of input–output analyses. For example, Munksgaard & Pedersen (2001) identify a foreign 'trade balance' in pollution as the difference between total emissions estimated on the basis of the PAP and CAP, or more simply, the difference between the pollution embodied in exports and the pollution embodied in imports. Particularly in the ecological footprint literature, where focus is accounting for emissions under the CAP, input–output analysis has become increasingly common as a technique to measure and allocate responsibility for emissions generation (for reviews, see Wiedmann *et al.*, 2007; and Wiedmann, 2009). As explained by Turner *et al.* (2007) this would seem a natural development, given that the focus of consumption-based measures such as the carbon footprint is to capture the *total* (direct plus indirect) resource use embodied in final consumption in an economy. Input–output analysis is based around a set of sectorally disaggregated economic accounts, where inputs to each industrial sector, and the subsequent uses of the output of those sectors, are separately identified. Therefore, by the use of straightforward mathematical routines, the interdependence of different activities can be quantified, and all direct, indirect and, where appropriate, induced resource use embodied within consumption can be tracked (Leontief, 1970; Miller & Blair, 2009).

Issues of jurisdiction and spatial focus

There are several issues that are not fully addressed in the existing input–output pollution accounting literature. One is that appropriate data are not commonly available for full CAP accounting (which, in a globalised economy would essentially require a world interregional input–output accounting framework) and commonly have to be estimated.

However, a more fundamental issue is that in moving between the extremes of PAP and CAP focus one may lose sight of the fact that PAP emissions are generated in the process of generating income/gross domestic product (GDP) in the producing economy. That is, the producing region/nation gains the benefits of value-added associated with its export production. Moreover, producers therein make decisions regarding production and associated polluting technology that determine emissions levels. It would seem, then, not straightforward that it is appropriate simply to allocate responsibility for pollution embodied in export production to trade partners where final consumers are located (Turner et al., 2011a). More generally, it may be that the extremes of the full PAP and CAP measures identified to date in the literature may not be appropriate for all pollutants or of practical policy interest. For example, for pollutants with local health impacts, it would seem clear that a focus on PAP emissions is crucial, regardless of responsibility concerns.

Another gap in the literature is that most applications to date have focused on national economies and international trade. However, with increasing decentralisation of responsibility for setting and/or achieving environmental and other sustainability objectives, it is appropriate to extend the accounting focus to sub-national regional economies and the pollution content of inter regional as well as international trade flows. One issue is whether devolved responsibilities are accompanied by the policy tools to meet these responsibilities. However, a more fundamental issue is that the spatial location of activities may benefit the nation in terms of meeting targets such as the PAP ones under agreements such as Kyoto. For example, within the UK access to low-carbon renewable resources for electricity generation is significant in Scotland relative to other regions. However, while this benefits the UK in terms of both its PAP and CAP positions, it punishes Scotland in terms of the simple metric of comparing PAP and CAP at the regional level (this is due to the fact that adopting cleaner technology in Scottish production reduces the pollution content of its exports but not its imports) and more specifically in terms of the inter-regional carbon trade balance within the UK (McGregor et al., 2008). Where devolved regions have responsibility for 'sustainability' issues, the location of pollution generation and pollution trade balances would seem to be an important, yet to date unconsidered, component of the UK devolution package.

Sub-national pollution accounting is important more generally. Notwithstanding what tend to be significant investments in transportation infrastructure that have reduced interregional transport costs in developed economies such as the United Kingdom and United States, it tends to be the case that sub-national regional economies are often be rather heterogeneous in structure. However, they are also likely to be highly interdependent. For example, the Midwest economies considered in the second case study in this paper (Wisconsin, Illinois, Indiana, Ohio and Michigan) are marked by a high degree of trade interdependence. On average between 30% and 40% of each state's trade is with the other four states (on both the import and the export sides). This generates a need to consider this collection of states as an important economic region. Moreover, given the level of interdependence, the economic–environmental linkages are likely to be different than in the rest of the United States, creating a need to explore this region in detail.

This paper attempts to address the issues identified above by considering the application of environmental input–output accounting techniques for different regional case studies and different types of air pollutant. The third section introduces the analytical environmental input–output framework and considers alternative treatments of the pollution embodied in trade flows. The fourth section applies input–output accounting techniques to the case study of carbon dioxide (CO_2) emissions generated in and/or attributable to the UK regions (focusing on the two-region case of Scotland and the rest of the UK (RUK)). Following this the paper turns its attention to a different geographical case with quite different policy concerns, focusing on the Midwest states of the United States and ammonia (NH_3) generation as an example of a non-GHG pollutant generated in a key trading sector of the regional economies therein (agriculture). The final section offers some conclusions and directions for future research.

The analytical environmental input–output framework

This section provides a technical exposition of the environmental input–output accounting method applied in the fourth and fifth sections. Readers are also referred to the non-technical overview and consideration of implications provided in Table 1.

The basic interregional environmental input–output framework applied to pollution generation

The interregional framework derived in Turner *et al.* (2007) is applied to demonstrate an analytical input–output method for enumerating the pollution content of trade flows. It begins with the standard, single region, Leontief inverse input–output equation (Leontief, 1970; Miller & Blair, 2009):

$$x = (I - A)^{-1}y \tag{1}$$

where **x** is an $N \times 1$ vector of gross outputs with elements x_i, where $i = 1, \ldots, N$, for each economic sector, i; and y is an $N \times 1$ vector of final demands with corresponding elements y_i. **A** is the technical coefficients matrix with elements a_{ij}, where $j = 1, \ldots, M$ and $M = N$. The **A**-matrix is derived from the input–output transactions matrix, where x_{ij} is the intermediate purchase of output in sector i as an input to production of output in sector j, X_j. Thus, each element of the **A**-matrix is formally defined as:

$$a_{ij} = x_{ij}/X_j \tag{2}$$

Table 1. Key aspects of different IO approaches for regional pollution analysis

	Factors included in analysis	Issues for environmental analysis
Direct single region – Equation (3)	* Domestic pollution (PAP) generation in target region	* Analysis entirely from a PAP perspective
Single region (Type 1) attribution to final demand – Equation (4)	* Domestic pollution (PAP) generation in target region * Direct and indirect (backward linkage/inter-industry) effects	* Attribution of PAP emissions to local and external consumers * No account of pollution embodied in imports
Full interregional 'footprint' analysis – Equations (6) and (7) at global level	* Actual (estimated) pollution generation supported by target region consumption (including interregional feedback effects) * Potential full CAP application	* Pollution attributed to local consumption demands dependent on production/polluting technology decisions that are not under the jurisdiction of regional policymakers (or consumers) * Extensive data requirements (requires world interregional IO tables, with economic and environmental data in IO format for all direct and indirect trade partners)
Limited interregional analysis using Equations (6) and (7) for target block with ROW trade estimated through Equation (10)	* Actual (estimated) pollution generation supported by target region consumption (neglecting interregional feedback effects) * Potential full CAP application	* Pollution attributed to local consumption demands dependent on production/polluting technology decisions that are not under the jurisdiction of regional policymakers (or consumers) * Less extensive data requirements than full interregional – for exogenous trade partners, only require data on imports from and direct pollution intensities for producers
Limited interregional analysis with ROW trade endogenised Equations (6) and (7)	* Domestic pollution(PAP) generation in target block, with CAP analysis among block members * Direct, indirect and import-induced effects * Allows more focus on choices regional policymakers have jurisdiction over	* Attribution of PAP emissions in block entirely to consumers in regions in block * Emissions generated in producing exports assumed to be required to facilitate imports * Pollution content of interregional trade measure, but no account of actual pollution embodied in imports from external

where $\mathbf{I}$ is the identity matrix. The $N \times N$ Leontief inverse is defined as $(\mathbf{I} - \mathbf{A})^{-1}$ with elements b_{ij}, describing the amount of output generated in each sector i for each unit of final demand for the output of sector j.

This standard input–output framework is augmented with a vector of output-pollution coefficients for a single pollutant, or a $(K \times N)$ matrix in the presence of $k = 1, \ldots K$ pollutants. Taking the multiple pollutant case, total pollution generation in production is defined as:

$$\mathbf{f^x} = \mathbf{\Phi x} \tag{3}$$

where $\mathbf{f^x}$ is a $K \times 1$ vector, with element f_k^x, where $k = 1, \ldots, K$, representing the physical amount of pollutant k generated within the economy through the production of the vector of gross outputs, $\mathbf{x}$. $\mathbf{\Phi}$ is a $K \times N$ matrix where element $\Phi_{k,i}$ is the amount of pollutant k per unit of gross output in sector i. In the analysis presented here, for simplicity it is assumed that $K = 1$, and k is a single pollutant (CO_2 in the UK case and NH_3 in the US Midwest case below). Where final consumers directly generate emissions, (3) may be extended to take account of this. See equation (7) for the interregional case. Here the authors proceed by focusing on production emissions for simplicity. However, the key point is that (3), or its extended variant including direct emissions by final consumers, will give the basic PAP account for the target region (Table 1).

Substituting equation (1) into equation (3) produces:

$$\mathbf{f^x} = \mathbf{\Phi}(I - A)^{-1}\mathbf{y} \tag{4}$$

This explains the $K \times 1$ vector of total pollutants generated in production in terms of the $N \times 1$ vector of final demand that is assumed to drive all activity in the economy.

Turner *et al.* (2007) extended this single-region framework to the case of two or more regions, $r = 1, \ldots, R$ and $R = S$, where R indicates producing and S indicates consuming regions. For simplicity of exposition, the framework is stated in terms of two regions, but it is straightforward to extend to the multi-region case. The final demand is presented as a matrix with separate elements for local final demand in region 1 for unit output produced in region 1 ($\mathbf{y}_{11}$) local (exogenous) final demand in region 2 for unit output produced in region 2 ($\mathbf{y}_{22}$). There are also direct exports to *exogenous* final demand in region 2 of unit output produced in region 1 ($\mathbf{y}_{12}$); and vice versa ($\mathbf{y}_{21}$). Exports to *endogenous* intermediate demand in region 2 for region 1 unit output given by $\mathbf{A}_{12}$, and vice versa. Thus, $\mathbf{x_{rs}}$ is an $N \times 1$ vector giving output of sectors in region r generated by the consumption demands (domestic and imports) of region s. Equation (1) can therefore be presented for the (two-region) interregional case as:

$$\begin{pmatrix} \mathbf{x}_{11} & \mathbf{x}_{12} \\ \mathbf{x}_{21} & \mathbf{x}_{22} \end{pmatrix} \begin{pmatrix} \mathbf{I} - \mathbf{A}_{11} & -\mathbf{A}_{12} \\ -\mathbf{A}_{21} & \mathbf{I} - \mathbf{A}_{22} \end{pmatrix}^{-1} \begin{pmatrix} \mathbf{y}_{11} & \mathbf{y}_{12} \\ \mathbf{y}_{21} & \mathbf{y}_{22} \end{pmatrix} \tag{5}$$

where $\mathbf{x_{rs}}$ is an $N \times 1$ vector giving output of sectors in region r generated by the consumption demands (domestic and imports) of region s.

Each region has $i = j = 1,\ldots,N$ production sectors where each sector i produces only one commodity j. Sub-matrices $\mathbf{A_{rs}}$ therefore contain elements a_{ij}^{rs}, describing the transactions between commodity production sector i in producing region r and consuming sector j in consuming region s, for each unit output of sector j in region s. $[\mathbf{I} - \mathbf{A}]^{-1}$ is the partitioned interregional Leontief inverse (multiplier matrix). Using a similar notation to that used for the single-region model, b_{ij}^{rs} is the output required in industry i in region r per monetary unit of final demand for industry j in regions s. Thus, by partitioning the $\mathbf{A}$-matrix so as to identify intermediate inputs production in the home and other region(s), and by separating the $\mathbf{y}$ vector final demand into commodities produced in the home and other region, it is therefore possible to identify how exogenous demand in one region affects activity in each other region.

As for the single-region case, this framework can be extended to consider the issue of pollution spillovers between regions. Equation (5) is augmented with $(1 \times N)$ vectors of output-pollution coefficients for a single pollutant $\mathbf{\Phi_r^x}$ (again, this could be replaced by a $(K \times N)$ matrix in the presence of $k = 1,\ldots,K$ pollutants). Each output-pollution vector shows the direct pollution intensity of output in each production sector i for an individual region, r:

$$\begin{pmatrix} f_{11}^x & f_{12}^x \\ f_{21}^x & f_{22}^x \end{pmatrix} = \begin{pmatrix} \mathbf{\Phi_1^x} & 0 \\ 0 & \mathbf{\Phi_2^x} \end{pmatrix} \begin{pmatrix} \mathbf{I} - \mathbf{A_{11}} & -\mathbf{A_{12}} \\ -\mathbf{A_{21}} & \mathbf{I} - \mathbf{A_{22}} \end{pmatrix}^{-1} \begin{pmatrix} \mathbf{y_{11}} & \mathbf{y_{12}} \\ \mathbf{y_{21}} & \mathbf{y_{22}} \end{pmatrix}$$
$$= \begin{pmatrix} \mathbf{\Phi_1^x L_{11} y_{11}} + \mathbf{\Phi_1^x L_{12} y_{21}} & \mathbf{\Phi_1^x L_{11} y_{12}} + \mathbf{\Phi_1^x L_{12} y_{22}} \\ \mathbf{\Phi_2^x L_{21} y_{11}} + \mathbf{\Phi_2^x L_{22} y_{21}} & \mathbf{\Phi_2^x L_{21} y_{12}} + \mathbf{\Phi_2^x L_{22} y_{22}} \end{pmatrix}$$

(6)

The first subscript on each element of equation (6) identifies the producing region, r, and the second identifies the consuming region, s. $\mathbf{L_{rs}}$ is that sub-matrix of the partitioned Leontief inverse that gives the total impact on the output in the producing sectors in region r per unit of final demand for output in region s. f_{rs}^x is a scalar representing the amount of pollution generated in production activities in region r to support region s final demand. Thus, $\mathbf{f_{rr}^x}$ indicates the amount of pollution used in production activities in region r to support final demand in region r. $\mathbf{f_{sr}^x}$ is the amount of pollution used in production activities in region s to support final demand in region r, and so on.

If final consumers also directly generate emissions, these are incorporated through a $1 \times Z$ vector, $\mathbf{\Phi_r^y}$, of expenditure-pollution coefficients for each final consumption group z in region r. Each element Φ_z^y describes the physical amount of pollution that is directly generated per monetary unit of final expenditure. Generally, one final consumption group, households (hh), generate direct emissions, so $z = hh =$

1, and this emissions generation only takes place in the home region. Therefore:

$$\begin{pmatrix} f_1^{hh} & 0 \\ 0 & f_2^{hh} \end{pmatrix} = \begin{pmatrix} \Phi_1^{hh} & 0 \\ 0 & \Phi_2^{hh} \end{pmatrix} \begin{pmatrix} y_1^{hh} & 0 \\ 0 & y_2^{hh} \end{pmatrix} \tag{7}$$

By summing the partitioned matrices in equations (6) and (7), it is possible to measure all emissions in regions $r = 1,\ldots,R$ that are attributable to the final consumption demand in each region for the outputs of the other region(s). For example, total emissions generated in region 1 (emissions generated within region 1 under PAP) are found by summing along the first row of each **f**-matrix so that:

$$f_1^x = f_{11}^x + f_{12}^x + f_1^{hh} \tag{8}$$

And total emissions in both regions 1 and 2 that are attributable to region 1 final consumption demand (emissions under CAP) are found by summing down the first column of each **f**-matrix so that:

$$f_1^y = f_{11}^x + f_{21}^x + f_1^{hh} \tag{9}$$

In accordance with Munksgaard & Pedersen's (2001) CAP principle, region 1's pollution 'trade balance' with region 2 is calculated as the difference between equations (8) and (9), and the corresponding calculations for region 2 are carried out using the second row and column of the **f** matrices in (6) and (7). This means that the pollution 'trade balance' (PTB) is given by:

$$PTB = f_{rs}^x - f_{sr}^x \tag{10}$$

In the framework outlined above, the treatment of interregional trade between regions 1 and 2 is in accordance with the CAP, such that total pollution generation within the geo-political area covered by these regions is attributed to consumption demand therein. Thus, if data are available to define a system that encompasses all (direct and indirect) trade partners of all regions $r = 1,\ldots,R = S$ included, equation (9) would give the full pollution 'footprint' (e.g. the carbon footprint) of region 1.

Within the framework outlined above, it is also possible to decompose the structure of the pollution supported by different types of final consumption in each region (household and government consumption and capital formation, as well as any external demands from outwith the system—see below). It is also possible to decompose in terms of the specific sectoral outputs that are consumed by different types of consumers (i.e. pollution generation in sector i in region r supported by final consumption group z in region s).

Treatment of external (international) trade

If all regions/countries that the target region trades (directly or indirectly) with are not accounted for in the system in (6) a decision must be made on how any external trade is dealt with. Generally, input–output tables for any target region, r, will record export demand from other regions and/or the rest of the world (ROW) for each sectoral output, i, as a column or columns within the $\mathbf{Y}$-matrix. However, this will usually only identify the destination region (e.g. the Scottish input–output tables report two columns for total exports from each production sector to RUK and ROW respectively), but not the using/consuming sector/final consumption group therein. Moreover, there will generally be a corresponding row reporting imports from the other region(s), but only in terms of the total value of imports to each sector i and final consumption group, z, not the sectoral commodity outputs that are used/consumed.

Where data are available to identify or estimate the full $\mathbf{A}_{rs}$ and $\mathbf{y}_{rs}$ matrices/vectors, and corresponding pollution intensity vectors, $\mathbf{\Phi}_y^x$ (and, where appropriate, $\mathbf{\Phi}_r^y$) for trade partners (as has been possible here for the sub-national cases of Scotland and RUK, and the five Midwest states and the rest of United States (RUS)) equation (6) can be populated and estimates under both PAP and CAP and the corresponding pollution trade balance can be determined using equations (7) to (10). However, as explained by Turner *et al.* (2007) and highlighted in Table 1, in the presence of extensive global trade, consideration of a full global CAP measure one is likely to effectively require a world interregional input–output framework, identifying all of the target region's direct and indirect trade partners and differences in production and carbon emitting technologies therein (also Andrew *et al.*, 2009).

Moreover, even if it were possible to identify such a database to analyse the resource requirements of final consumption in the region of interest, the second section has raised issues relating to jurisdiction (also Turner *et al.*, 2011a; Peters & Hertwich, 2008) and policy concern with regard to different types of pollutant that may make it appropriate and/or desirable to adopt an alternative approach. Turner *et al.* (2011a) propose an alternative extension of the system in (4) that focuses on unidirectional trade flows (imports to the target region only, dropping exports from final consumption, $\mathbf{y}$, in order to focus on domestic consumption only).

$$f_r^y = \mathbf{\Phi}_w^x[\mathbf{I} - (\mathbf{A}_{rr} + \mathbf{A}_{sr})]^{-1}(\mathbf{y}_{rr} + \mathbf{y}_{sr}) + \mathbf{\Phi}_r^y \mathbf{y}_{rr}^* \qquad (11)$$

where * denotes a transpose vector, and with total final consumption by each final consumption group, z, generally collapsing to households as in (7) above.

The system in (11) only requires data on imports in input–output matrix format in addition to the single region framework. Data are used on the commodities as the source of UK imports and corresponding CO_2 emissions produced by colleagues at the Organisation for Economic Co-operation and Development (OECD) (Turner *et al.*, 2011b) to populate the $1 \times N$ vector $\mathbf{\Phi}_w^x$, of weighted direct pollution intensities for each commodity output i. The weights attached to the direct carbon intensity of output in each producing country, r, are given by the share of commodity output i from region/country r in total region s use of commodity output.

Of course (11) does not capture interregional feedback effects or multiplier effects in the region(s) that the target region imports from. Where this is desirable for some subset of regions that the target region trades with (for example, for a group of regions within a single national economy), it is possible to utilise both (6) and (11), using the former to consider intra-national trade and the latter for international trade. A mixed approach is adopted for the case of Scotland and RUK below.

Another issue raised above is that policy concern with regard to different types of pollutants may mean that the full interregional approach applied to all trade as in (6) is not appropriate, desirable or useful. For example, McGregor *et al.* (2008) argue that where national pollution targets under the Kyoto Protocol and/or Copenhagen Accord relate to emissions generation within national borders, it may be appropriate to consider intra-national trade under the CAP using (6) but apply the PAP to trade with ROW. This may involve one of two approaches. First, (6) could be calculated including exports to ROW from each region, r, within that region's pollution account as part of $\mathbf{y_{rr}}$ and $\mathbf{f_{rr}^{x}}$, but with no consideration of the pollution implications of imports. Second, (6) could be adjusted to consider the domestic (rather than global) pollution implications of the region's import requirements by considering the pollution involved in export production that finances imports through the nation's current account.

McGregor *et al.* (2008) argue in favour of the latter adjustment and develop (6) to give what they refer to as a trade endogenised linear attribution system (TELAS). This is for a national (UK) pollution accounting framework where all regions $r = 1$, ..., R ($= S$) are regions within that national economy. The TELAS approach involves endogenising trade in much the same way as household final consumption is endogenised in a standard Type II analysis (Miller & Blair, 2009). Instead of including international export demand for each region within the $\mathbf{Y}$ final demand matrix, an additional national production sector is created in the partitioned $\mathbf{A}$-matrix, a Trade sector, t, which produces goods for trade to facilitate the imports required in the national economy as a whole. The row entries for each sector j in each (consuming) region s are that sector's imports from ROW as a share of the total input/output, X_j. The additional column entries are the outputs that must be produced for export to ROW via the Trade sector, t, by each (producing) sector i per unit of total imports required in the UK economy as a whole (intermediate and final consumption), which is the output of the Trade sector. The pollution intensity of the output of the new national UK Trade sector is equal to zero, as no emissions are directly generated here (emissions directly generated in producing output for export demand are generated in the producing sectors and are, therefore, embodied indirectly in intermediate sales to the new Trade sector).

Thus, when (6) is calculated for the extended system with trade endogenised, this means that each individual (production or consumption) sector that imports from ROW will be attributed the pollution embodied in the share of total national domestic export production required to finance these imports. Because the PAP is applied at the national level, no attempt is made to estimate the actual pollution embodied in imports Instead, the TELAS approach focuses on pollution generated *within* the national economy to support consumption therein.

The fifth section considers the application of the TELAS approach in the context of the generation of acid rain precursors, the emissions of which, while impacting over relatively large geographical areas, are generally considered in terms of direct PAP emissions within a national context. It considers the case of emissions of a single non-GHG air pollutant, ammonia (NH_3), generated mainly as a result of agricultural production activity, the reduction of which is considered to be particularly challenging (Kaiser, 2001) in the context of the US Midwest. Here the paper extends on the standard PAP to consider what demand patterns within the United States (including demand for imported goods and services) drive levels of agricultural production and pollution therein.

Illustrative case study 1: Accounting for CO_2 generation in and attributable to UK regions under different accounting principles

The first case study is to examine UK CO_2 generation in the accounting year of 2004, broken down by six production sectors (an aggregation of the 123 Standard Industrial Classification (SIC)-defined input–output categories reported in UK national and regional input–output accounting[1]) and two regions, Scotland and RUK.[2] In the case of Scotland, the paper draws on the 2004 Scottish input–output tables,[3] along with experimental data on imports from RUK and ROW in input–output format and sectoral CO_2 intensity data provided by the Scottish government input–output team. In the case of the UK, where input–output data are not published in the required analytical format (symmetric tables in producer/basic prices), the paper draws on data published by Wiedmann *et al.* (2008) to construct a UK industry-by-industry analytical input–output table for 2004.[4] This is augmented with data on imports from ROW in input–output format constructed by Wiedmann *et al.* (2008) and carbon intensity data from the UK environmental accounts.[5] The interregional framework to populate equation (6) is then constructed in the manner outlined for 1999 in McGregor *et al.* (2008). RUK sectoral emissions are simply taken as the difference between UK and Scottish emissions at the six-sector level and the output-pollution coefficients derived by dividing through by activity levels (Figures 1 and 2).

Figure 1 shows that Scottish emissions intensities are not greatly different from UK averages except in the case of Energy (which includes gas and electricity distribution). Here the CO_2 intensity of the Scottish sector is significantly lower than the UK average due to the greater use of renewable electricity generation technologies. This is important information. As discussed above, a strategy to lower UK PAP emissions, and that element of UK CAP that is generated at home, may be to locate production in regional areas where lower emissions intensities may be achieved while still generating the same level of output and GDP. In Extraction, Quarrying, Construction and Water Supply activities, on the other hand, the Scottish CO_2 intensity is markedly

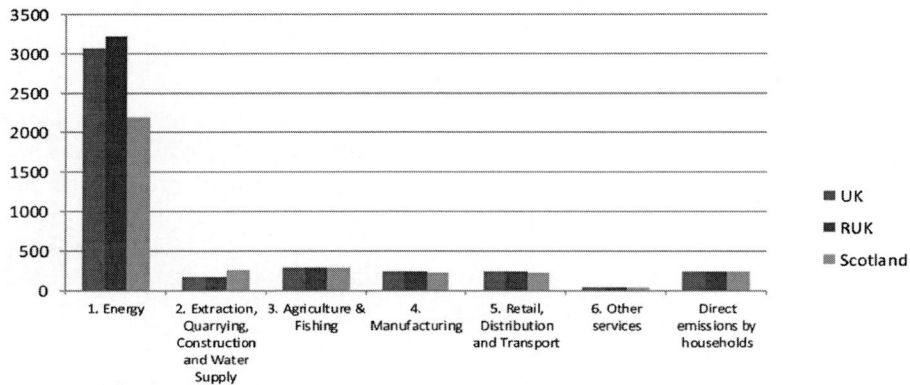

Figure 1. Output-CO_2 coefficients (production sectors) and hourshold final expenditure-CO_2 pollution coefficients for UK, RUK and Scotland 2004 - tonnes of CO_2 (equivalent) per £1m output/expenditure

higher. However, this is largely due to the differential composition of activity in the aggregate sector.

Figure 2 shows the sectoral generation of CO_2 at the regional and national level, where just over 8% of UK emissions are directly generated in Scotland (a similar proportion to the GDP generated in Scotland in the same accounting period). This is the direct PAP account.

However, one can get a better understanding of the regional structure of CO_2 generation, and of the extent of CO_2 'trade' between Scotland and RUK, by estimating equation (6) where the **A**-matrix is a $2N \times 2N$ or a 12×12 (with six sectors in each region) partitioned matrix. Here the outputs of UK production sectors are treated as endogenous while the partitioned matrix **Y** of final consumption demands includes export to both producers and consumers in ROW. That is, one begins with a conventional Type I (Miller & Blair, 2009) open economy attribution analysis to understand the drivers of UK regional PAP generation (what is accounted for under agreements such as the Kyoto Protocol and Copenhagen Accord).

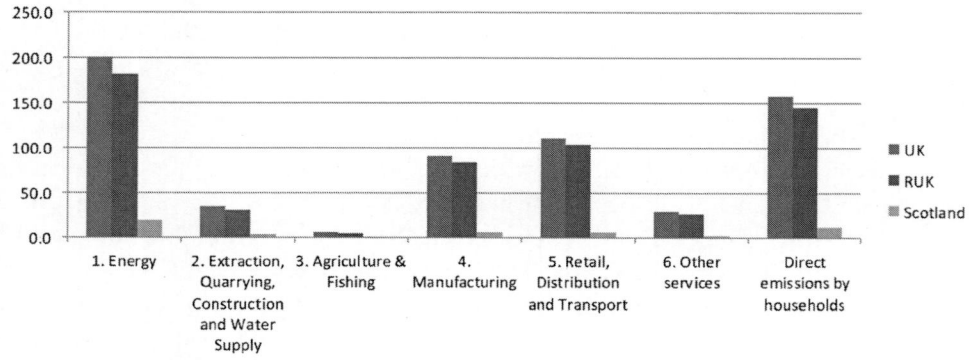

Figure 2. Direct CO_2 (millions of tonnes of CO_2 equivalent) Emissions Generated in UK, RUK and Scotland in 2004

Table 2 shows the scale of the CO_2 'trade' (or 'spillovers') that occurs between Scotland and RUK in the accounting year. This shows that just over 30% (16.2 million tonnes of CO_2 measured as CO_2 equivalent) of the total CO_2 generated to support conventional Scottish final demand expenditures is generated in RUK (i.e. not in Scotland). A similar proportion of CO_2 generated in Scotland is to support, directly or indirectly, RUK final demand (15.5 million tonnes). Also note that Scottish exports to ROW require 2.6 million tonnes of CO_2 to be generated in RUK as a result of the indirect impacts of the production of intermediate inputs.

The sectoral distribution of direct CO_2 generation in each region is shown in the final column of Table 2. Along each row, it can be seen how this breaks down by final consumption demand in each region (including both domestic regional demands and also ROW export demand for each region's output). The largest share of CO_2 embodied in trade flows between the two regions is embodied in trade in Energy sector outputs, which is not surprising given the pollution intensity of this type of production. While Energy production is less CO_2 intensive in Scotland (Figure 1), emissions embodied in production to support RUK demands, 8.9 million tonnes (including the 1.7 million tonnes supported by ROW demand for RUK outputs), still accounts for just under 17% of total CO_2 emissions in Scotland. Next to Energy, trade in aggregate Manufacturing outputs accounts for the most important component of CO_2 embodied in interregional trade flows (21% of CO_2 embodied in Scotland to RUK trade and 28% in the other direction).

The bottom of Table 2 shows that there is a negative CO_2 trade balance for Scotland—here labelled as the interregional pollution trade balance (IPTB)—as opposed to Scotland's net regional balance (RPTB) (Table 4)—implying that the pollution generated in Scotland by production supporting RUK final demands is less than the pollution generated in RUK by production supporting Scottish final demands. However, the Scottish CO_2 trade deficit (−0.7 million tonnes) is relatively small, accounting for just 1.25% of total CO_2 generated in Scotland. Moreover, Turner *et al.* (2011b) identify how this 'deficit' relationship is driven by the fact that cleaner electricity generation technology in Scotland (incorporated in the Energy sector here) reduces the level of CO_2 embodied in exports to RUK, rather than Scottish imports being particularly CO_2 intensive.

However, policy attention at both the regional and national level in the UK has increasingly begun to focus on whether full CAP measures may be adopted instead of or as well as PAP measures. If one considers CO_2 emissions attributable to, for example, Scottish regional consumption demands under a full CAP, CO_2 embodied in exports (to RUK and/or to ROW) should not be included. Rather, if policy and public interest shifts to measuring the 'carbon footprint', this requires focus on emissions required anywhere in the world to support regional demands. As discussed in the second section, it also implies a shift in focus from sources of both pollution and income generation to the implications of consumption decisions. Moreover, in moving to consider the global CO_2 impacts of consumption decisions in the UK and her regions, one begins to take account of pollution generated as a result of

Table 2. The CO$_2$ Trade Balance Between Scotland and RUK (tonnes, millions) - Type I Interregional Input-Output attribution analysis

Pollution generated in:	Pollution supported by:				
	Scottish HH/GOVT/ CAPITAL	Scot-ROW	RUK HH/GOVT/ CAPITAL	RUK-ROW	Total regional emissions of CO$_2$
Scotland					
1. Energy	7.7 (14.5%)	2.7 (5.1%)	7.2 (13.6%)	1.7 (3.3%)	19.2 (36.5%)
2. Extraction, Quarrying, Construction and Water Supply	2.6 (4.8%)	0.3 (0.6%)	1.0 (1.8%)	0.2 (0.4%)	4.0 (7.6%)
3. Agriculture & Fishing	0.3 (0.6%)	0.2 (0.5%)	0.4 (0.7%)	0.1 (0.2%)	1.0 (2.0%)
4. Manufacturing	0.8 (1.6%)	2.3 (4.4%)	2.5 (4.7%)	0.7 (1.3%)	6.3 (11.9%)
5. Retail, Distribution and Transport	5.0 (9.4%)	0.7 (1.3%)	1.1 (2.2%)	0.2 (0.4%)	7.0 (13.2%)
6. Other services	1.7 (3.3%)	0.1 (0.3%)	0.5 (0.9%)	0.1 (0.1%)	2.4 (4.6%)
Direct CO$_2$ generation by Scottish households	12.8 (24.3%)				12.8 (24.3%)
Total CO$_2$ generation in Scotland	30.9 (58.5%)	6.4 (12.1%)	12.5 (23.8%)	3.0 (5.6%)	52.8 (100%)
RUK					
1. Energy	6.5 (1.1%)	1.7 (0.3%)	119.8 (20.7%)	54.2 (9.4%)	182.2 (31.5%)
2. Extraction, Quarrying, Construction and Water Supply	0.6 (0.1%)	0.1 (0.01%)	25.7 (4.4%)	4.6 (0.8%)	30.9 (5.3%)
3. Agriculture & Fishing	0.2 (0.03%)	0.1 (0.01%)	4.0 (0.7%)	1.2 (0.2%)	5.5 (0.9%)
4. Manufacturing	4.0 (0.7%)	0.5 (0.1%)	40.8 (7.1%)	38.8 (6.7%)	84.1 (14.6%)

(*Continued*)

Table 2 Continued

Pollution generated in:	Pollution supported by: Scottish HH/GOVT/ CAPITAL	Scot-ROW	RUK HH/GOVT/ CAPITAL	RUK-ROW	Total regional emissions of CO_2
5. Retail, Distribution and Transport	1.9 (0.3%)	0.2 (0.04%)	84.5 (14.6%)	17.7 (3.1%)	104.3 (18.0%)
6. Other services	0.4 (0.1%)	0.1 (0.01%)	21.9 (3.8%)	4.2 (0.7%)	26.6 (4.6%)
Direct CO_2 generation by RUK households			144.6 (25.0%)		144.6 (25.0%)
Total CO_2 generation in RUK	13.5 (2.3%)	2.6 (0.5%)	441.5 (76.3%)	120.7 (20.9%)	578.3 (100%)
Total (UK) CO_2 emissions supported by	44.4 (7.0%)	9.0 (1.4%)	454.0 (71.9%)	123.6 (19.6%)	631.1 (100%)
Interregional pollution trade balance (IPTB)					
Scot pollution supported by RUK final demand		15.5 (=12.5 + 3.0)			
RUK pollution supported by Scot final demand		16.2 (=13.5 + 2.6)			
Scotland's CO_2 trade deficit with RUK		−0.7			

Table 3. Regional carbon footprint estimates for Scotland and RUK (2004) – broken down by composition of Scottish, RUK and ROW commodities directly or indirectly consumed

	Carbon footprint by commodity source in each region (tonnes, millions)			Total carbon footprint by commodity source
	Scottish commodities	RUK commodities	ROW commodities	
Scotland				
1. Energy	7.7 (10.1%)	6.5 (8.5%)	10.7 (14.1%)	24.8 (32.7%)
2. Extraction, Quarrying, Construction and Water Supply	2.6 (3.4%)	0.6 (0.7%)	0.6 (0.8%)	3.8 (4.9%)
3. Agriculture & Fishing	0.3 (0.4%)	0.2 (0.3%)	0.9 (1.2%)	1.5 (1.9%)
4. Manufacturing	0.8 (1.1%)	4.0 (5.2%)	5.9 (7.8%)	10.7 (14.1%)
5. Retail, Distribution and Transport	5.0 (6.5%)	1.9 (2.5%)	12.9 (17.0%)	19.8 (26.0%)
6. Other services	1.7 (2.3%)	0.4 (0.5%)	0.5 (0.7%)	2.7 (3.5%)
Indirect CO_2 embodied in consumption	18.1 (23.8%)	13.5 (17.8%)	31.6 (41.6%)	63.2 (83.2%)
Direct CO_2 generation by final consumers	12.8 (16.8%)			12.8 (16.8%)
Total carbon footprint (by regional source)	30.9 (40.7%)	13.5 (17.8%)	31.6 (41.6%)	76.0 (100%)

	Carbon footprint by commodity source in each region (tonnes, millions)			Total carbon footprint by commodity source
	Scottish commodities	RUK commodities	ROW commodities	
RUK				
1. Energy	7.2 (1.0%)	119.8 (16.5%)	70.8 (9.7%)	197.8 (27.2%)
2. Extraction, Quarrying, Construction and Water Supply	1.0 (0.1%)	25.7 (3.5%)	9.4 (1.3%)	36.0 (5.0%)
3. Agriculture & Fishing	0.4 (0.05%)	4.0 (0.6%)	11.2 (1.5%)	15.6 (2.1%)
4. Manufacturing	2.5 (0.3%)	40.8 (5.6%)	90.9 (12.5%)	134.2 (18.5%)
5. Retail, Distribution and Transport	1.1 (0.2%)	84.5 (11.6%)	82.9 (11.4%)	168.6 (23.2%)
6. Other services	0.5 (0.1%)	21.9 (3.0%)	8.1 (1.1%)	30.5 (4.2%)
Indirect CO_2 embodied in consumption	12.5 (1.7%)	296.8 (40.8%)	273.4 (37.6%)	582.8 (80.1%)
Direct CO_2 generation by final consumers		144.6 (19.9%)		144.6 (19.9%)
Total carbon footprint (by regional source)	12.5 (1.7%)	441.5 (60.7%)	273.4 (37.6%)	727.4 (100%)

Table 4. Summary of composition Scottish PAP, CAP and full regional pollution trade balance RPTB

Scotland:

	PAP	Scottish Demand	RUK Demand	ROW demand for RUK outputs	ROW demand for Scottish outputs
Scotland	52.8 (100%)	30.9 (58.5%)	12.5 (23.8%)	3.0 (5.6%)	6.4 (12.1%)
	CAP	Scottish production	RUK production		ROW production
	76.0 (100%)	30.9 (40.7%)	13.5 (17.8%)		31.6 (41.6%)
RPTB (PAP-CAP)	−23.2	0	−1.0	3.0	−25.2

RUK:

	PAP	RUK Demand	Scottish Demand	ROW demand for Scottish outputs	ROW demand for RUK outputs
RUK	578.3 (100%)	441.5 (76.3%)	13.5 (2.3%)	2.6 (0.5%)	120.7 (20.9%)
	CAP	RUK production	Scottish production		ROW production
	727.4 (100%)	441.5 (60.7%)	12.5 (1.7%)		273.4 (37.6%)
RPTB (PAP-CAP)	−149.1	0	1.0	2.6	−152.7

production decisions in other countries, where UK regional and national policy-makers have limited or no policy jurisdiction.

In the third highlighted row of Table 3 equation (6) has been used to calculate emissions within the UK required to support Scottish consumption (44.4 million tonnes) and RUK consumption (454 million tonnes) respectively. However, no account has been taken of emissions embodied in imports from ROW. Table 3 uses equation (11) to estimate this.

The first thing to note is that the CAP figures in Table 3 are considerably higher than the PAP figures in Table 2: the Scottish footprint (76 million tonnes) is 44% larger than its domestic PAP emissions (52.8 million tonnes), while the UK footprint (727.4 million tonnes) is 26% higher (578.3 million tonnes).[6] While Scotland directly generates 40% of its carbon footprint, in RUK this is higher at 61%. Thus, it can be observed that a smaller, more open region has less direct control over its CAP account. The commodity composition of the two regional footprints is quite varied, with a greater share of the Scottish carbon footprint (33% compared with 27% in RUK) originating in Energy production and a greater share of the RUK carbon footprint (18.5% compared with 14% in Scotland) in Manufacturing. Direct household emissions are also more important in the RUK case (20% of the total footprint) than in the Scottish case (17%).

Table 4 summarises each region's gross pollution trade balance (RPTB) in terms of the conventional difference between PAP and CAP, breaking this down in terms of its regional and international trade components.

Despite the concerns raised above, it may be argued that CO_2 is a GHG and climate change is a global problem so it would seem appropriate to develop accounting frameworks to examine CAP emissions at least alongside PAP measures. Moreover, CAP addresses the concern that PAP emissions may be reduced by importing 'dirty' goods and services (though a singular focus on CAP may reduce incentives to 'clean up' domestic polluting technology). The accounting framework developed here facilitates additional attention to the pattern of domestic emissions and pollution embodied in trade between regions, which may prove useful in considering the most environmentally efficient location of production within a nation, but also the implications for regional policy-makers in setting and meeting emissions targets.

Illustrative case study 2: Accounting for and attributing responsibility for NH_3 emissions generation in agricultural production in the US Midwest

In the case of non-GHG pollutants which have more localised impacts, on the other hand, alternative measures may be more appropriate and informative. For example, the United States has tended to focus on voluntary reduction of emissions of non-CO_2 GHGs (methane and fluorocarbon emissions) and is not a signatory to the Kyoto Protocol or Copenhagen Accord. Here much of the federal environmental regulation and policy has focused on air quality within the United States.[7] Emissions of acid rain precursors (sulphur dioxide, SO_2 and nitrous oxides, NO_x) have received particular focus under the Clear Air Act Amendments (CAAA) as well as the Acid

Deposition Control Program of 1990 and the George W. Bush administration's Clear Skies Initiative of 2003. Pollution spillover effects are given attention in terms of trans-boundary air pollution issues between the United States and Canada in the 1991 Air Quality Agreement, but this also focuses primarily on acid rain precursors, as does the 2005 Clean Air Interstate Rule issued by the US Environmental Protection Agency (USEPA) (which caps emissions of SO_2 and NO_x in the 28 Eastern US states including the five Midwest states considered here and in the District of Columbia, which are particularly affected by acidic deposition).

Another air pollutant of significant policy concern in the United States is NH_3. NH_3 emissions may actually neutralise acid rain, or even make it alkaline, but may cause soil acidification through nitrification. As with emissions of the main acid rain precursors (SO_2 and NO_x), formation of secondary particulates from NH_3 emissions may react with organic compounds to contribute to ozone formation, causing vegetation, material and health damage as well as affecting visibility (Menz & Seip, 2004). While NH_3 emissions are generated through transport and other industrial activities (for example, the US's Comprehensive Environmental Response Compensation Liability Act, CERCLA, of 1980 focuses on emissions from chemical and petroleum industries), the main sources are agricultural through livestock operations and the use of fertilisers. This paper focuses on NH_3 here to demonstrate how input–output analysis may be used to understand the structural nature of emissions from a particular sectoral source.

The paper uses the ten-sector, six-region input–output tables for the Midwest and RUS derived using the method of Jackson et al. (2006) from 2007 IMPLAN[8] US interregional input–output and commodity flow data. The output-NH_3 coefficients to populate the $(1 \times N)$ vector $\mathbf{\Phi_r^x}$ for each of the five Midwest states accounted for (Illinois, Indiana, Ohio, Michigan and Wisconsin) and RUS are derived from research carried out at the Regional Economics Applications Laboratory, University of Illinois, and funded by the USEPA STAR programme (for more detailed methodology, see Tao et al., 2007). The USEPA STAR-funded research also identifies emissions intensities for a range of other non-GHG pollutants.

The NH_3 intensities (tonnes per US$1 million sectoral output) for the most NH_3-intensive activity, Agriculture, Forestry and Fisheries are shown in Figure 3 for the United States as a whole, the five Midwest states and for RUS. It is agricultural production, particularly involving the use of fertilisers and/or livestock operations, that is the main source of NH_3 emissions. With an NH_3 intensity of just over 33 tonnes per US$1 million output, the agricultural activity in the state of Wisconsin is the most intensive in the production of this pollutant. While one cannot break out the data to this extent, it is the particular composition of agricultural activity in Wisconsin that explains this NH_3 intensity: the Wisconsin profile from the 2007 US Census of Agriculture[9] shows that Wisconsin is the second largest US producer of 'milk and other dairy products from cows', 'other animals and animal products' as livestock operations, and the largest producer of 'corn for silage', which involves the use of fertilisers. These activities are highly NH_3 intensive and make up a large proportion of the overall Agriculture, Forestry and Fisheries sector in Wisconsin.

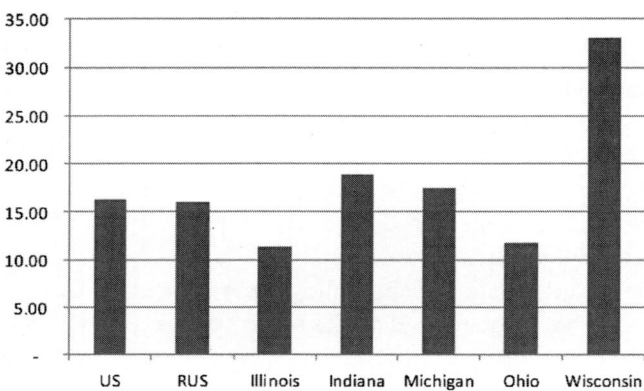

Figure 3. Output-NH$_3$ Coefficients for Agriculture, Forestry and Fisheries Sector for US, RUS and 5 Mid-West states. (tonnes of NH$_3$ per \$1m output/expenditure)

The first numerical column of Table 5 shows the physical amount (in tonnes) of NH$_3$ directly generated in Agriculture (the Agriculture, Forestry and Fisheries sector) in each state (the dominating source of all NH$_3$ emissions, as shown in the final column). Note that in all six regions identified at least 80% of NH$_3$ emissions are from agricultural sources and in the case of Wisconsin this rises to 98.4%. Policy tends to focus on the direct sources of these emissions. However, as in the case of CO$_2$ above, input–output methods can be used to understand the structure of the pollution generation problem in terms of the sources of demand driving these emissions. Therefore Table 5 applies equation (6) for the R = S = 6 region, N = 10 sector case to attribute NH$_3$ emissions to exogenous final demands originating in each of the five Midwest states, RUS and ROW.

Reading along the Wisconsin row of Table 5, the Type I input–output analysis shows that (in the accounting year of 2007) 34.3% of agricultural NH$_3$ emissions are attributable to own-region final consumption demands, 16.8% to final consumption in other Midwest states and 27.8% to other US.

Another interesting result is that a large share (19.4%) of agricultural NH$_3$ generated in Wisconsin is supported by consumption demand outside the United States. On the other hand, when reading down the second to last column of Table 5, it can be seen that Wisconsin has the lowest share of its NH$_3$ emissions supported by external ROW demands. Nonetheless, Table 6 shows that Wisconsin overall import requirement relative to its export production is also relatively low. That is, it exports far more (with a value of US\$37.5 billion in 2007, of which 27.4% was from the Agriculture, Forestry and Fisheries sector, the highest share across the states in Table 7) than it required to finance its own imports (which were valued at US\$13.7 billion). It is in this context that the TELAS approach discussed above may prove useful. Table 7 repeats the analysis using equation (6), but this time it endogenises trade with ROW under the assumption that exports (inputs to a new US-level 'Trade' sector) are produced in order to finance or facilitate imports (the output of the new Trade sector) to support final consumption demands within the United States.

Table 5. NH3 Emissions generated in Mid-West regions and RUS attributed under a Type I input-output analysis

Pollution generated in:	Total NH$_3$ emissions (tonnes)	Pollution supported by: (percentage of regional total NH$_3$ emissions)							
		I FD	J FD	M FD	O FD	W FD	RUS FD	Export ROW	Total
Illinois	173,650								
Agriculture	139,328	9.7%	2.8%	2.3%	2.1%	1.2%	27.9%	34.2%	80.2%
Indiana	173,962								
Agriculture	166,173	7.1%	14.5%	3.7%	6.0%	0.9%	27.7%	35.7%	95.5%
Michigan	136,409								
Agriculture	125,636	3.5%	2.3%	34.8%	5.8%	2.0%	22.1%	21.4%	92.1%
Ohio	110,127								
Agriculture	97,000	2.2%	2.8%	4.3%	29.5%	0.4%	22.0%	26.8%	88.1%
Wisconsin	347,632								
Agriculture	341,936	6.1%	1.3%	7.8%	1.7%	34.3%	27.8%	19.4%	98.4%
RUS	5,542,104								
Agriculture	5,187,270	3.0%	1.2%	1.6%	2.1%	0.7%	64.3%	20.7%	93.6%
Total US NH$_3$ emissions	6,483,885	3.8%	1.8%	2.9%	3.0%	2.7%	63.4%	22.4%	100.0%

Table 6.   State level trade balances with ROW in 2007 ($million)

	Exports to ROW	Of which agricultural	Imports from ROW	Trade surplus with ROW	Ratio imports to exports
Illinois	73,809	5.7%	25,108	48,702	0.34
Indiana	45,291	19.3%	22,109	23,182	0.49
Michigan	55,080	13.0%	36,588	18,491	0.66
Ohio	69,355	11.8%	28,875	40,480	0.42
Wisconsin	37,508	27.4%	13,707	23,802	0.37
RUS	1,267,269	25.5%	1,133,136	134,133	0.89

Table 7. NH_3 Emissions generated in Mid-West regions and RUS attributed under a TELAS input-output analysis

Pollution generated in:	Total NH_3 emissions (tonnes)	Pollution supported by: (percentage of regional total NH_3 emissions)						
		I FD	J FD	M FD	O FD	W FD	RUS FD	Total
Illinois	173,650							
Agriculture	139,328	10.8%	3.5%	3.6%	3.2%	1.8%	57.3%	80.2%
Indiana	173,962							
Agriculture	166,173	8.2%	15.2%	5.1%	7.1%	1.5%	58.4%	95.5%
Michigan	136,409							
Agriculture	125,636	4.2%	2.8%	35.7%	6.5%	2.4%	40.6%	92.1%
Ohio	110,127							
Agriculture	97,000	3.0%	3.4%	5.4%	30.3%	0.9%	45.1%	88.1%
Wisconsin	347,632							
Agriculture	341,936	6.7%	1.7%	8.5%	2.3%	34.6%	44.6%	98.4%
RUS	5,542,104							
Agriculture	5,187,270	3.6%	1.6%	2.4%	2.8%	1.1%	82.1%	93.6%
Total US NH_3 emissions	6,483,885	4.6%	2.2%	3.8%	3.7%	3.0%	82.7%	100.0%

The statistics shown in Table 6 reflect the fact that all five Midwest states have relatively low import to export ratios. In Table 7 the NH_3 emissions in each region that are attributed to ROW in the Type I analysis (in Table 5) are essentially reallocated *pro rata* to the sectors and final consumers in each region that import. From this viewpoint, the cost of imports, in both economic and environmental terms (with the latter focusing here on NH_3 emissions), are those associated with the exports that production sectors in each region have to provide to pay for US imports. Again taking the example of Wisconsin agriculture row, if one compares the results in Tables 5 and 7, observe that while there are small increases in the percentage of emissions attributable to each of the Midwest states, the largest increase (from 27.8% to 44.6%) is in NH_3 emissions generated in Wisconsin that are attributable to consumption demand in RUS. That is, almost half of NH_3 emissions produced in this state are required to support final consumption activity (including imports) in the non-Midwest states. Similar increases and magnitudes for NH_3 emissions supported by RUS consumption are observed across all five Midwest states.

In the case of a more localised pollutant like NH_3, where environmental impacts such as soil acidification will be felt locally (and impact on the future economic costs of carrying out production activities), demand path analysis of the type that is facilitated particularly by the TELAS approach would seem to be of potential usefulness to policy-makers. The core finding reflected in Table 7 is that the Midwest states bearing a net environmental cost (in terms of the impacts of a damaging agricultural pollutant) in producing to support consumption activity in the United States as a whole. It may be quite rightly argued that the Midwest states are producing and trading based on their comparative (and resource) advantage in agriculture. However, the findings raise questions to policy-makers in terms of who should bear the costs of environmental damage in particular areas of the United States to support consumption demands in the nation as a whole.

Discussion and conclusions

This paper uses input–output accounting methods to consider issues of pollution attribution from different accounting perspectives that may be useful to regional and national policy-makers in considering climate change and other environmental policy objectives.

Input–output accounting techniques are already accepted, particularly in the economic systems and ecological economics literature, as providing appropriate methods to track pollution embodied in complex economic interactions and supply chains. However, to date most applications have focused on case studies for national economies and on international trade. The research reported here takes a more sub-national/ regional level focus and considers what may be achieved using currently available/ accessible data to provide analysis and results that may be useful in different jurisdictional contexts and with respect to different pollutants.

The main arguments and findings are as follows. First, both PAP and CAP may be measured in the same input–output accounting framework and it is argued that both should be computed and reported, particularly in informing policy. A singular focus on PAP or CAP has implications in terms of incentives to import 'dirty goods' or to focus on 'cleaning up' consumption rather than production. Moreover, policy-makers may be limited in the extent to which they can control one or the other and the information offered by either PAP or CAP alone is limited. Examination of both using a common analytical framework permits a better understanding of the structure of pollution problems, whether these are global or local in impact, though the precise techniques utilised may vary depending on this (see below).

The paper also proposed a sub-national focus in analysing elements and drivers of PAP and CAP measures. Again, doing so permits greater insight into the structure of pollution problems within a nation, particularly where the composition of production and/or consumption activity is heterogeneous, but with a high degree of interdependence between regions. This heterogeneity and interdependence between regions of a national economy may also mean that the conventional extremes of PAP, CAP and associated pollution trade balances give a misleading and incomplete picture. This is apparent in the UK analysis above, where Scotland's use of low-carbon renewable electricity generation technology reduces the pollution content of its exports, with great impact on its regional PAP than CAP.

However, through the second case study it was also argued that in the case of pollutants with more localised impacts than GHGs, full CAP measures are not really relevant or useful, with a necessity to focus on emissions generation at a local level under PAP. However, it is proposed that demand path analysis may be useful, using an adapted input–output technique to focus within a national PAP framework, particularly where relevant environmental regulations are set at this level.

Acknowledgements

The research reported in this paper was funded through the Economic and Social Research Council (ESRC) Climate Change Leadership Fellow project 'Investigating the Pollution Content of Trade Flows and the Importance of Environmental Trade Balances' (ESRC Ref. No. RES-066-27-0029), based at the universities of Stirling and Strathclyde. It is also supported by the Shetland Northern Isles New Energy Solutions (NINES) project. The authors acknowledge the invaluable assistance of the Input–Output team at the Office of the Chief Economic Adviser, Scottish Government, colleagues at the Stockholm Environment Institute (York, UK), and Norihiko Yamano at the Organisation for Economic Co-operation and Development (OECD) in constructing data used in the UK application.

Notes

1. Generally input–output accounting is carried out at a greater degree of sectoral detail than that reported in the case studies here. This paper opts for a higher level of aggregation in the illustrative case studies primarily to report disaggregated results and explain these to the reader. However, there is a more practical motivation in that the quality of the experimental Scottish government data on Scottish sectoral emissions is questionable at a greater degree of sectoral disaggregation.
2. Scotland is the only region of the UK for which official input–output tables are published by a government agency.
3. For Scottish input–output tables for 1998-2007, see http://www.scotland.gov.uk/input-output/.
4. For the UK analytical input–output table for 2004, see http://www.strath.ac.uk/fraser/research/2004ukindustry-byindustryanalyticalinput-outputtables/.
5. For the UK environmental accounts for 2004 in input–output format, see http://www.statistics.gov.uk/statbase/Product.asp?vlnk=14883&image.x=14&image.y=9/.
6. Throughout the analysis here UK Environmental Accounts data are used that include emissions from UK aviation and shipping.
7. For example, the 1970 Clean Air Act (CAA) and the subsequent 1977 and 1990 Clear Air Act Amendments (CAAA) set deadlines for compliance with nationwide ambient air quality standards (NAAQS).

8. See http://implan.com/V4/Index.php/.
9. For the US 2007 Census of Agriculture, see http://www.agcensus.usda.gov/.

References

Andrew, R., Peters, G. & Lennox, J. (2009) Approximation and regional aggregation in multi-regional input–output analysis, *Economic Systems Research*, 21, 311–35.

Bastianoni, S., Pulselli, F. M. & Tiezzi, E. (2004) The problem of assigning responsibility for greenhouse gas emissions, *Ecological Economics*, 49, 253–57.

Ferng, J. J. (2003) Allocating the responsibility of CO_2 over-emissions from the perspectives of benefit principle and ecological deficit, *Ecological Economics*, 46, 121–141.

Jackson, R. W., Schwarm, W. R., Okuyama, Y. & Islam, S. (2006) A method for constructing commodity by industry flow matrices, *Annals of Regional Science*, 40, 909–920.

Kaiser, J. (2001) The other global pollutant: nitrogen proves tough to curb, *Science*, 294, 268–269.

Kondo, Y., Moriguchi, Y. & Shimizu, H. (1998) CO_2 emissions in Japan. Influences of imports and exports, *Applied Energy*, 59, 163–174.

Leontief, W. (1970) Environmental repercussions and the economic structure: an input–output approach, *Review of Economic and Statistics*, 52, 262–277.

McGregor, P., Swales, J. K. & Turner, K. (2008) The CO_2 trade balance between Scotland and the rest of the UK: performing a multi-regional environmental input–output analysis with limited data, *Ecological Economics*, 66, 662–673.

Menz, F. C. & Seip, H. M. (2004) Acid rain in Europe and the United States: an update, *Environmental Science and Policy*, 7, 253–265.

Miller, R. & Blair, P. (2009) *Input–output analysis: foundations and extensions* (2nd edn) (Cambridge, Cambridge University Press).

Mongelli, I., Tassielli, G. & Notarnicola, B. (2006) Global warming agreements, international trade and energy/carbon embodiments: an input–output approach to the Italian case, *Energy Policy*, 34, 88–100.

Munksgaard, J. & Pedersen, K. A. (2001) CO_2 accounts for open economies: producer or consumer responsibility?, *Energy Policy*, 29, 327–334.

Peters, G. & Hertwich, E. (2008) CO_2 embodied in international trade with implications for global climate policy, *Environmental Science and Technology*, 42, 1401–1407.

Tao, Z., Williams, A., Donaghy, K. & Hewings, G. (2007) A socio-economic method for estimating future air pollutant emissions—Chicago case study, *Atmospheric Environment*, 41, 5398–5409.

Turner, K., Lenzen, M., Wiedmann, T. & Barrett, J. (2007) Examining the global environmental impact of regional consumption activities—Part 1: A technical note on combining input–output and ecological footprint analysis, *Ecological Economics*, 62, 37–44.

Turner, K., Munday, M., Jensen, C. D. & McIntyre, S. (2011a) Incorporating jurisdiction issues into regional carbon accounts under production and consumption accounting principles, *Environment and Planning A*, 24, 722–741.

Turner, K., Yamano, N., Druckman, A., Ha, S. J., De Fence, J., McIntyre, S. & Munday, M. (2011b) An input–output carbon accounting tool with carbon footprints estimates for the UK and Scotland, *Fraser of Allander Institute Economic Commentary* (Special edn), January, 6–20.

Wiedmann, T. (2009) Editorial: Carbon footprint and input–output analysis: an introduction, *Economic Systems Research*, 21, 175–186.

Wiedmann, T., Lenzen, M., Barrett, J. & Turner, K. (2007) Examining the global environmental impact of regional consumption activities—Part 2: Review of input–output models for the assessment of environmental impacts embodied in trade, *Ecological Economics*, 61, 15–26.

Wiedmann, T., Wood, R., Lenzen, M., Minx, J., Guan, D. & Barrett, J. (2008) *Development of an embedded carbon emissions indicator*, Report to the UK Department for Environment Food and Rural Affairs, Stockholm Environment Institute & Centre for Integrated Sustainability Analysis, University of Sydney.

Wyckoff, A. W. & Roop, J. M. (1994) The embodiment of carbon in imports of manufactured products: implications for international agreements on greenhouse gas emissions, *Energy Policy*, 22, 187–194.

Decentralising energy: comparing the drivers and influencers of projects led by public, private, community and third sector actors

Bouke Wiersma and Patrick Devine-Wright

Department of Geography, University of Exeter, Exeter, UK

The potential contribution of decentralised energy (DE) to the low carbon transition has received increasing policy and scholarly attention. However, a predominant emphasis upon community-led initiatives has overlooked the potential of alternative configurations, in particular projects led by public, private and professional third sector actors. To address this gap, a comparative case study analysis was undertaken based upon in-depth interviews with key actors in nine UK DE projects, scrutinising cross-sectoral patterns in underlying project drivers and factors influencing project evolution. Findings indicate that drivers are highly diverse, vary by sector and are predominantly local, with addressing poverty predominant. Key influencers identified were funding, levels of trust and stakeholder representations of energy users. The results indicate that policy and academic emphases on community-led DE overlook other successful and diverse configurations that can contribute to the low carbon transition.

1. Introduction

The IPCC's most recent reports have highlighted the substantial and multifaceted challenges posed by climate change (Intergovernmental Panel on Climate Change [IPCC], 2014). One important strategy in climate change mitigation is carbon emission reduction within the energy supply sector, which accounts for 38% of UK CO_2 emissions (Department of Energy and Climate Change [DECC], 2014), and 35% globally (IPCC, 2014). Therefore, the IPCC calls for 'a fundamental transformation of the energy supply system, including the long-term substitution of unabated fossil fuel conversion technologies by low-GHG (greenhouse gas) alternatives' (p. 4).

Energy systems in developed countries are highly centralised, involving large-scale power stations and a 'national grid' infrastructure to distribute electricity to centres of demand (DECC, 2011; Watson & Devine-Wright, 2011). To achieve decarbonisation as part of a climate change mitigation strategy, a more decentralised energy (DE) system has been proposed by both advocacy groups (Adams & Berry, 2008; Greenpeace, 2005; Hathway, 2010; Julian & Dobson, 2012) and academics (Alanne & Saari, 2005; Mulugetta, Jackson, & Van der Horst, 2010). Indeed, it touches upon wider debates within the social sciences around the appropriate scale for a transition to low carbon energy (Catney et al., 2014; Hargreaves, Hielscher, Seyfang, & Smith, 2013).

From a technical point of view, DE refers to the supply of electricity and heat generated on or near the site where it is used (cf. BERR, 2008). However, it is the best understood as a multi-dimensional concept, encompassing technical, financial, political and social aspects (Watson & Devine-Wright, 2011). DE includes a range of technologies that utilise renewable energy sources such as solar (hot water and photovoltaic systems), hydro, wind and biomass (e.g. wood pellet boilers). It can be deployed at a range of scales, from the household to the neighbourhood or urban level (Aiken, 2012)[1] and may be accompanied by demand-side measures designed to reduce or shift energy consumption.

A range of benefits have been claimed for DE, including reduced carbon emissions (Bouffard & Kirschen, 2008; Greenpeace, 2005), enhanced energy security (Alanne & Saari, 2005; Bouffard & Kirschen, 2008), behavioural changes leading to lower energy consumption (Bergman & Eyre, 2011), enhanced acceptability of energy developments (Devine-Wright, Walker, Hunter, High, & Evans, 2007; Sauter & Watson, 2007) and providing local empowerment and financial opportunities (Alanne & Saari, 2005; Hathway, 2010; Walker, 2008). The final point suggests the potential value of a decentralised alternative in enabling local resilience and localised climate change mitigation and adaptation. However, the disconnect between these apparent benefits of DE, and the continued dominance of a centralised, fossil fuel-based energy system, begs questions around what drives the emergence of DE initiatives and how they subsequently evolve – a better understanding of which should enable the design of policies that galvanise the more widespread emergence of similar initiatives. More particularly, as a consequence of policy and academic focus on community energy projects – outlined in the next section – there is a distinct lack of understanding of the development and evolution of DE initiatives beyond the community sector. This prevents a more complete understanding of the potential contribution of the full variety of local actors engaged in DE, climate change and behaviour change actions in achieving a low carbon transition.

Therefore, there are two pertinent questions this paper aims to address:

(1) What drives the emergence of DE initiatives, and what influencers shape their evolution?
(2) To what extent do these drivers and influencers differ, depending upon the sector of the instigating actor (public, private, third, community)?

Systemically, Watson and Devine-Wright (2011) described climate change as a key factor shaping the recent emergence of DE, alongside other factors such as energy security, maturing technologies and a broader shift towards the engagement of energy users and grassroots action. Several other studies have begun to investigate the drivers of DE by examining specific, typically community-led initiatives. These have revealed the importance of economic (e.g. employment) and environmental (e.g. climate change) drivers, and also what Bomberg and McEwen (2012) term 'symbolic resources': seeing community energy projects as a vehicle to increase local autonomy, cohesion or community identity. Adams and Berry (2008) reviewed five community energy projects commissioned by a UK rural regeneration charity. They concluded that these initiatives were driven by a range of issues, from local needs (reliable electricity supply, regeneration of local economy) to wider environmental motives linked to climate change (reducing carbon emissions, reducing dependence on fossil fuels). Seyfang, Park, and Smith (2013) showed that community-led DE initiatives are driven by multiple objectives (an average of 8 per case in their survey analysis). Of these, the most frequent were financial motivations (e.g. 96% of groups, for example, saving money on bills, generating income for the community), but environmental (88% of groups), social (73%), political (73%) and infrastructural (68%) drivers were also prominent. Studies in the Netherlands found similar results, in that reducing local dependence upon large energy companies was a key driver of DE (Boon, 2012; Chin-A-Fo, 2012, cited by Arentsen & Bellekom, 2014).

This emerging literature suggests that the drivers of DE are multiple and diverse, and likely to vary from project to project. However, the research to date has been skewed towards investigating 'bottom-up' projects led by community actors. Whilst this has been successful in deepening our understanding of grassroots initiatives, it has overlooked the dynamics of DE projects led by other stakeholders, namely public sector bodies (e.g. local authorities), private companies and third sector organisations (e.g. NGOs). This is an important omission, given that recent research found that only 36% of a total of 182 UK urban energy projects were led by the third sector (conceptualised as both volunteer-led community groups and professional third sector organisations that employ paid staff members), the remainder being public and private sector projects (Turcu & Rydin, 2012). If the potential of a more decentralised pathway is to be fully understood, research needs to move beyond a narrow focus on community-led DE towards a broader spectrum of initiatives emerging in this field.

One study that has looked beyond communities to include stakeholders from other sectors suggests that in fact drivers of DE are likely to be different across sectors (Allen, Sheate, & Diaz-Chavez, 2012). Based upon interviews with public, private and third/community sector stakeholders in a community DE project in the Lake District National Park, UK, the authors found that stakeholders' ideas of the principal drivers of community renewable energy differed – public sector stakeholders judged various 'top-down', national energy policies as key drivers, while community sector stakeholders were driven by the social benefits associated with community DE schemes. A second study that has specifically investigated differences between community-owned and private sector-owned local energy infrastructures has suggested that community-owned developments may also be more acceptable (Warren & McFadyen, 2010). All of this suggests that the dominant framing of DE as a community-based exercise may in fact be overlooking and simplifying the complexities behind the development and success of DE development that stems from the involvement of parties other than community groups.

In summary, while research has explored the drivers of community-led DE, manifesting multiple economic, environmental and social goals, similar evidence is lacking for DE initiatives led by other sectors. Furthermore, the ways in which initiatives across the entire DE pathway evolve, and are influenced by which factors, have received little attention. Further investigation of this evolution is important to better understand the ways in which initiatives develop, stall or change, if the sector is to be guided and assisted to achieve its carbon reduction potential. Therefore, this study aims to enhance understanding of these issues for all forms of DE initiatives using a comparative analytical approach.

2. Methodology

A comparative case study design was chosen to investigate nine DE projects (Table 1), which were selected on the basis of diversity. Projects were led by stakeholders from across the community, private, public and third sectors. Here, we distinguish between community and third sectors with the community sector understood to consist of voluntary groups, and the third sector of professional organisations like NGOs. Case study selection also aimed to reflect diversity of spatial location, selecting areas in England, Wales, Scotland and Northern Ireland, and involving multiple interventions ranging from heat and power generation using biomass district heating or photovoltaic panels to demand reduction measures like improved insulation.

Employing a broadly interpretative and constructivist epistemology, a mixed-method approach was adopted consisting of the analysis of secondary data (e.g. project reports, website materials) and 33 in-depth, semi-structured interviews with all available key project

Table 1. Description of the nine case study projects, ordered by sector of the leading stakeholder(s).

Sector	Case study	Leading stakeholder(s)	Focus of project interventions
Community sector projects	Glencraig (Northern Ireland)	Camphill Community Glencraig & Camphill Community Trust Northern Ireland (both community sector)	Heat generation
		This community installed a biomass-fuelled district heating system in 2011 to connect 21 mixed-use buildings, using low-grade waste biomass from their own land, local sawmills and tree surgeons, the Northern Ireland Railway, and a range of other local and regional sources	
	Sustainable Moseley (England)	Sustainable Moseley (community sector), St Mary's Church, Hamza Mosque (both third sector)	Electricity generation and Demand reduction
		Set-up in 2007, Sustainable Moseley is a volunteer-based community group concerned with supporting the local community and reducing carbon emissions. In addition to resident engagement activities, solar electricity systems were installed on 4 community buildings, and 18 local homes received energy improvements, funded by a grant from British Gas Green Streets	
Private sector projects	Newport (Wales)	Solutia & Wind Prospect group (both private sector)	Electricity generation
		On the site of chemical company Solutia, two grid-connected 2.5 MW wind turbines were installed in 2009 by developers Wind Prospect Group, funded by pension funds investment and income from the UK government's Renewables Obligation	
	Zero Carbon Homes (England)	Scottish and Southern Energy (SSE), BRE, PRP Architects (all private sector)	Heat and electricity generation and demand reduction
		In this research project, energy provider SSE funded the construction of 10 Zero Carbon Homes with attached energy centre, which were completed in 2010 and used to test new technologies (such as biomass district heating and heat pumps) and gain experience with the concept of Zero Carbon Homes from an energy company's perspective	

(Continued)

Table 1. Continued.

Sector	Case study		Leading stakeholder(s)	Focus of project interventions
Public sector projects	Riverside Dene (England)	This regeneration project (2010–2012) refurbished five 1960s' social housing tower blocks, combined with the installation of a biomass district heating system fuelled by wood pellets to replace electric storage heaters in the flats, funded by the a national government grant	Newcastle City Council & Your Homes Newcastle (both public sector)	Heat generation and demand reduction
	Wandle Valley Low Carbon Zone (England)	Instigated by the Greater London Authority, this two-year project finished early 2012, having aimed to achieve an area-wide 20% reduction in carbon emissions in the London Borough of Merton, through resident engagement at schools, 'Green Doctor' home energy audits, and social housing retrofits	London Borough of Merton Council (public sector), Groundwork London (third sector) and Sustainable Merton (community sector)	Electricity generation & demand reduction
Third sector projects	Energy Neighbourhoods (England)	Instigated by the Belgian environmental federation BBL replicating their 'Klimaatwijken' initiative, and funded by a grant from the EU under the Intelligent Energy programme, groups of around eight households formed 'Energy Neighbourhoods' competing to save as much as energy as possible by behavioural changes during the 2008–2009 winter	Severn Wye Energy Agency (third sector)	Demand reduction
	Renewable Heritage (Scotland)	This demonstration project, which finished in 2009, installed solar hot water systems into 49 tenement flats in 7 listed Georgian housing blocks in a heritage area, funded by national and private grants	Changeworks, Lister Housing Cooperative, Edinburgh World Heritage (all third sector), Edinburgh City Council (public sector)	Heat generation
	Shimmer (England)	Focusing on those in fuel poverty, and funded by the Technology Strategy Board and the Feed-in Tariff, in 2010, households were provided with free solar panels and a laptop with access to a personalised interactive energy management platform	London Rebuilding Society, Homezone (both third sector), Energy Saving Trust (quasi public)	Electricity generation demand reduction

stakeholders (e.g. project managers).[2] The use of interviews allowed for the expression of beliefs and opinions by the individuals who were directly involved with project management and implementation, capturing the multiple interpretations that might be held about the dynamics, drivers and influencers of the case study projects. The interview protocol was designed to capture the complexity of the case studies and therefore covered general elements (e.g. how and when did the project begin?), governance aspects (e.g. what is your organisation's role in the project?), finances (e.g. how was the project funded?), social aspects (e.g. to what extent have those affected by the project participated in planning and design?) and technology (e.g. which technologies have been used in this project?).

Interviews were audio-recorded, fully transcribed and subjected to thematic analysis (Braun & Clarke, 2006; Joffe, 2011) using the software NVivo. This resulted in a coding frame comprising 20 top-level and 18 sub-codes, which were both theoretically driven and inductively grounded in the data. The coding frame was checked for inter-researcher reliability by independently coding two entire interviews by two researchers, which revealed high levels of agreement and clarified and refined code definitions. To facilitate comparisons across the case studies, the coded data were summarised using a number of matrices (based on the key themes emerging from the thematic analysis), containing all the coded data per theme ordered per case. This enabled iterative, interpretative analysis that investigated cross-case patterns and compared these internally and also with other themes. Furthermore, patterns and commonalities emerging from the themes were analysed in relation to various typologies of the nine projects, in particular, focusing on differences between the community, private, public and professional third sector.

3. Results

The first section outlines the various driving factors behind project instigation, looking at the motivations mentioned by stakeholders for instigating or joining a project. The second section then discusses three factors that were represented by stakeholders as having significantly influenced the evolution of each project.

3.1. *Drivers of DE initiatives*

Analysis of the interview data revealed a rich and complex depiction of the drivers of DE initiatives. These were multiple – seven key drivers were identified in total, and each project was associated with at least two of these. They also varied between and within cases – it was typical for specific initiatives to reveal several drivers, which were of mixed importance for each of the partners involved.

The most common discourse throughout the interviews viewed DE as a means to address local poverty through reducing energy bills, often achieved through channelling DE subsidies and grants. This was mentioned as a key driver in five of the nine projects, and especially in public and third sector-led projects. Stakeholders interpreted this theme in terms of alleviating fuel poverty, improving housing quality standards, health and well-being. For example, the Shimmer project involved a partnership between three third sector stakeholders and interviews with each revealed the multiplicity of project drivers across stakeholders. For Dan[3] (Homezone), Shimmer was an attempt to reduce poverty:

'We wanted to find poor people. I think I, and the others, have been shocked by what we've found, because when you go into a house that's freezing, you've got four kids and they're all hungry, and the rest of it, you really want to do what you can. I would say it's driven us on.'

For Tim (London Rebuilding Society), Shimmer was driven by a concern for social equity and inclusion:

> 'What I want to see is low carbon energy production and secure energy production and also equitable access to it so in terms of fees and charges so that's one particular area where decentralised can mean, can be better … if that meant there was greater equity, more equitable charging tariffs and access to energy.'

For another stakeholder involved in this project (Energy Saving Trust), the driver for participating was purely environmental, and the meeting of specific policy targets around decarbonisation in particular:

> Terry, Energy Saving Trust (Shimmer): 'we want to save carbon is our main mission, to save carbon, that's why we were set up (…) the reason why projects like Shimmer have taken place, which is the advent of smart metering which is to enable the UK to effectively monitor and reduce its CO_2, we wouldn't be doing this if we didn't have CO_2 targets.'

This was in fact the only mention of national carbon targets as a driving factor across the nine projects – even though many were at least partly funded through schemes that aim to reduce carbon emissions (e.g. Renewables Obligation). In only two other projects were carbon targets, in both cases local rather than national targets, talked about as driving the project onwards. In both instances, these targets were mentioned by Local Authority stakeholders, who positioned them as rolled into the organisation's broader reputational and performance objectives:

> Jacob, Newcastle City Council (Riverside Dene): 'For the last two years Newcastle have been awarded this Most Sustainable City in the UK award which they're very, very proud of (…). It does a lot to try and tackle carbon reduction. So there's lots – and that's one of the big drivers for the council: the carbon reduction target.'

More broadly, stakeholders talked about DE as a way of caring for the environment, raising awareness of energy issues and fostering self-sufficient living. This theme was present in interviews across all four projects with strong involvement from community groups (Energy Neighbourhoods, Glencraig, Sustainable Moseley and Wandle Valley), and was absent from public and private sector-led projects. Environmental drivers were represented by some in explicit contrast to financial drivers:

> Ben, St Mary's Church (Sustainable Moseley): 'The original idea was to make a visible contribution to energy awareness and renewable energy generation, that was what motivated the thing. In fact the feed in tariff was not very high then (...) – the main incentive for people to do it now is the feed in tariff but then it wasn't.'

Financial benefits were described as a key motive for instigating DE in both private and one of the community projects, which referred to short-term, direct economic benefits like lower electricity rates (Newport) and switching to cheaper fuels (Glencraig) as well as long-term benefits like ensuring long-term profitability of the instigating stakeholder:

> Kate, SSE (Zero Carbon Homes): 'We wanted to see how zero carbon homes would impact our business (...) we wanted to identify the occupants' needs in these houses to see if they would be any different to our current sort of customers (…) so that we could provide energy in the future to a different type of customer in a low carbon world.'

This emphasis upon the economic benefits of DE for stakeholders complemented the social driver of reducing local poverty already remarked upon. Moreover, the interviews revealed the value associated with adopting low carbon energy technologies, in particular to enhance the public image or reputation of participating stakeholders. This was mentioned as a project motivator by three instigating stakeholders across the third (Renewable Heritage), public (Wandle Valley) and private (Newport) sectors, but was consistently represented as a secondary rather than a key driver:

> Vincent, Solutia (Newport): 'Well the key benefit is the money. Secondary things are (...) the business benefits of the sort of green sort of credentials that they provide. Yes, there is a bit of, you know, "Solutia, they've got two wind turbines".'

Given that many of the projects were funded as demonstration projects, it is unsurprising that researching and demonstrating were represented as important drivers across all sectors except the community sector. The research varyingly served to ensure profitability (Zero Carbon Homes), dissemination of lessons learnt (Renewable Heritage, Wandle Valley) and to inform local government planning guidance:

> Jane, Edinburgh City Council (Renewable Heritage): 'And it was a real piece of best practice work for us. Very much we needed an example to be able to show people that it could work. And we'd been refusing solar panels on listed buildings for a while, (...) with no real reason for it.'

In this particular case, there was a strong motivation to ensure that planning applications for renewable energy on listed buildings were consented or rejected on evidence-based grounds, learning lessons that could be shared with other areas with listed buildings:

> Jane, Edinburgh City Council (Renewable Heritage): 'And that was the reason why we got involved with the project, because we were really keen to see some evidence, and be able to – from a policy point of view – give that to our case officers who determine the applications, to say "no this isn't a problem".'

Drivers of urban DE were often represented as entangled within wider urban renewal, such as improving housing quality and improving neighbourhood character. For example, the Renewable Heritage project was said to be driven by the large amount of 'hard to treat' pre-1919 housing in Edinburgh and Scotland, while the Riverside Dene project was seen as a response to an urgent and multifaceted need to address poor housing quality in the local area:

> Liz, Newcastle City Council (Riverside Dene): 'I mean the accommodation by everybody's account wasn't up to standard, you know, you got wind about your windows, a very poor heating system, electric storage heaters. It was costing a huge amount of money for people – it just wasn't satisfactory accommodation at all.'

3.1.1. *Comparing drivers across the case studies*

Table 2 illustrates the multiple drivers that were observed in the interviews across the nine case studies and the differences across the four sectors. Whilst it is notable that discourses around reducing poverty and care for the environment were the most common, they were absent in the private sector discourse, which in turn most commonly talked about financial and reputational drivers. While carbon targets only played a limited role, broader environmental drivers were widespread, although strikingly in about half of the projects other drivers than environmental ones were more apparent in driving the project's emergence. Furthermore, in these cases at least, the community sector seems less driven by financial and research-oriented drivers than some of the other sectors. The table illustrates that the majority of drivers were of a local nature across all sectors (see also

Table 2. Discourses of drivers per case study and sector (black cells indicate that the driver was talked about by stakeholders in each case study; a white cell indicates that it was not)

Sector of leading stakeholder	Public		Private		Third			Commun.	
Drivers of instigating stakeholder(s) per project (ordered by frequency)	RD	WV	Ne	ZC	EN	RH	Sh	Gl	SM
Local drivers									
Reducing poverty and social exclusion	■	■				■			■
Improving local housing stock	■								
Benefitting stakeholder financially			■	■				■	
Improving stakeholder reputation		■				■			
Meeting carbon reduction targets	■						■		
Non-local drivers									
Caring for the environment		■		■	■		■	■	
Researching and demonstrating					■	■			

Devine-Wright & Wiersma, 2013). Discourses that were largely absent from the interviews included local empowerment and a more democratic energy system, trying to transition into a different type of energy system, and actively challenging powerful incumbents (like large energy companies). Although one project (Shimmer) was represented as reducing local social exclusion and vulnerability to a perceived impending 'energy gap' between supply and demand at the national level, this was one of the very few arguments made in relation to the wider energy system. The general absence of more systemic drivers from the interviewees suggests that these more structural, ideological aspects may not have been the most important drivers for these nine projects, even though commonly talked about in academic literature (Bomberg & McEwen, 2012) and advocacy documents (Hathway, 2010) around community-led DE.

3.2. *Influencers on DE initiatives*

Interview data suggested that the projects were co-shaped by many different influencing factors. This section explores three of these (funding, trust and stakeholder representations) which were influential across the nine projects, and helpful in understanding differences between projects led by the four sectors.

3.2.1. *Funding as both a facilitating and inhibiting influence upon DE*

While both private sector projects were funded through private investments (e.g. pension funds), the remaining seven projects received a one-off grant from parties outside the project management (e.g. national government). This had a number of repercussions that were absent in the private sector projects, including the need to comply with demands set by the funder to limit projects in scale (e.g. the number of neighbourhoods allowed to participate in Energy Neighbourhoods), location of implementation (Shimmer was tied to East London) and technology choice:

> Sophie, Sustainable Moseley (Sustainable Moseley): 'We also want to do a wind turbine because we had planned this for Green Streets, but we think that British Gas had a bit of a photovoltaic agenda – which is pretty clear that they do. So they said that they wouldn't be able to do it because we were too near houses in the allotments but we kind of think that's not true.'

Several projects were rushed due to grant spending deadlines, which complicated project implementation:

> David, Camphill Community Glencraig (Glencraig): 'The other thing was that of course this whole pressure of being in doubt if we could go ahead or not and then having actually through that planning delay nearly three months robbed of our time to do it in, and then get the whole project in practice rolling. That was very tight, being able to think of every detail.'

In some cases, projects were set up in very short periods of time in response to a sudden potential funding opportunity – projects in this category (Energy Neighbourhoods and Wandle Valley) were often indicated to experience teething problems due to a lack of existing social relations in the locality:

> Margaret, Groundwork London (Wandle Valley): 'There wasn't any resident group set up when we started in the zone so there was no method for us to engage with the community, so I think the most productive projects are always the ones where people are very active and want to be active in their community when you're at the start, that you begin the project together and I think that's been our problem.'

However, a more positive effect for those organisations successful in obtaining grants was that the grant success catalysed the development of all organisations involved:

> Sophie (Sustainable Moseley): 'And so winning that [Green Streets funding] I think changed everything (...) that kind of made us a much more focused organisation and it made it in some ways easier to get people to stay along with it because there was always stuff to do.'

3.2.2. *Trust and engagement with local energy users*

Trust was indicated to play a highly variable role across the different sectors; in fact, stakeholders frequently talked about residents mistrusting energy companies or the government, while not-for-profit organisations, in contrast, indicated that they had benefitted from being trusted by local residents:

> Dan, London Rebuilding Society (Shimmer): 'What you're also doing at the same time, is building a relationship so they trust you so you can do things like automatically set their radiators only to come on when people are in the room (...) otherwise if you said to them, do you want your electricity company to control your appliances, they say, I'm not having that bloody electricity company whatever, they don't trust them you see, but they trust us and I think all those things will help us hit our goals.'

However, trust is not purely dependent on sectoral origin, as in particular community and third sector projects were represented as having benefitted from a long-standing engagement with local residents, which was said to enable a smooth project implementation – for example, in Glencraig, there already had been seven years of discussions about implementing some form of DE, which meant that no objectors were left when funding was finally secured:

> Samuel, Camphill Community Trust Northern Ireland (Glencraig): 'We were in the fortunate position that when [the funding] came we have had a seven year period where people voiced their opinion and after the seven years we were able to implement that very quick and effective because they had voiced all their opinions – at the end (...) there was not one single objection.'

It should also be noted that all nine projects were represented as having encountered almost no opposition, regardless of whether they were community, public, private or third sector-led.

3.2.3. *Stakeholder representations of local energy users*

Across the projects, stakeholders represented local publics that were affected by, or benefitted from, the DE installation as lacking in education, understanding, awareness and concern about energy and climate change, which is a not uncommon way of imagining energy users (Barnett, Burningham, Walker, & Cass, 2012). These deficit representations of publics co-shaped the nine case studies in varying ways, revealing differences across the four sectors. In particular, for the private sector projects, these representations were associated with an emphasis on techno-logical solutions and a lack of active engagement with socio-behavioural aspects of DE, attributed to 'normal' lifestyles:

> Michael, researcher and resident (Zero Carbon Homes): 'We are encouraged to lead a normal life because they want to have the result of the research from normal family life, not somebody who is not having a tv, not having a computer, not using extra light, just keeping temperature as low as poss-ible. No, no – people are encouraged to have a normal life.'

By contrast, for community and third sector-led projects, stakeholders described a strong desire to engage with local energy users in order to make them aware of the new DE system and reduce their energy use. Responsibility for using the technologies to save energy was not simply passed on to householders, but instead actively taken on board by stakeholders, who went out of their way to maximise project outcomes in order to achieve enduring behavioural change:

> Scott, Lister Housing Cooperative (Renewable Heritage): 'it's for their benefit so if you just say, if you give someone a toy and say here it's for you and you can have it at no cost and go and play with it, then half of them will play with it for half an hour and then ignore it. So we wanted people to be interested and excited about it so that they would continue to use it for the next twenty years.'

However, such efforts to effect behavioural changes often focused on the financial rather than the environmental benefits of DE technologies. This was justified by project stakeholders by repre-senting local residents as unable to understand DE technologies and its context (e.g. climate change), and requiring an adjusted message:

> Margaret, Groundwork London (Wandle Valley): 'So I think if you look at things from that approach of "if we can install this in your bathroom it's going to save you money because it will reduce the amount of water you're heating up" then that's the way (...). Because if he's going there talking about we really want to help you reduce the amount of CO_2 you're producing (...), they're not going to understand anything.'

In other words, only DE projects led by some sectors show signs of attempts to enable 'energy citizenship' (Devine-Wright, 2007) involving attempt to disrupt 'normal' disengagement from energy supply and use, indicating that this commonly cited benefit of DE cannot be presumed to be universally present across all DE initiatives across sectors.

4. Discussion

This study aimed to address a predominant focus upon community-led DE initiatives in the litera-ture and to better understand the dynamics of initiatives led by public, private, third and commu-nity sector stakeholders in terms of their underlying drivers and the influences on their development. The findings indicate that the drivers and influencing factors are multiple and diverse, both within and between cases, and vary substantially across sectors. This multiplicity of drivers and influencers resonates with the multidimensionality of DE described by Watson and Devine-Wright (2011). DE projects were frequently driven by objectives such as reducing poverty and caring for the environment. Carbon emission reduction targets played a comparatively minor role, despite their predominance in government framings of DE (DECC, 2011). The diverse discourses invoked by stakeholders when discussing the instigation of the projects shows how, across its entire, highly diverse spectrum, DE is interpreted as much more than merely a means of local energy provision and should be more widely recognised as such. Though many of the seven discourses of drivers related to the potential benefits of DE are identified in the literature (see Alanne & Saari, 2005; Greenpeace, 2005; Hathway, 2010), others identified previously in relation to community energy projects, such as a desire for local empowerment and transitioning away from the current centralised energy system (Seyfang et al., 2013; Bomberg & McEwen, 2012), were less prominent. This suggests that these ideological concepts are less resonant with DE stakeholders outside the community sector, as DE in those sectors instead seems to be driven by more practical concerns such as poor-quality housing stock.

The diversity of discourses suggests that the application and evolution of DE is markedly different across these four sectors. Reducing poverty and caring for the environment were predominant discourses amongst third and community sector stakeholders, while private sector stakeholders mainly talked about financial gains. Only public sector stakeholders were driven by policies (e.g. local carbon targets or the UK Decent Homes Standard), supporting the previous research (Allen et al., 2012). Private sector stakeholders talked about regulatory incentives rather than policy; in contrast, community and third sector project stakeholders talked about neither policy nor regulations as driving their projects. In other words, community and third sector stakeholders invoked discourses that related to benefitting others and the environment, whereas public and private sector projects were more often driven by benefitting the project stakeholders themselves, predominantly by complying with regulations or benefitting financially.

These sectoral differences can be interpreted in the light of recent research that has shown that public support for energy system change, including the increased use of renewable energy, is contingent upon values such as social justice, equity and autonomy (Butler, Parkhill, & Pidgeon, 2013). Taking this finding more broadly, the differences found in the drivers of projects led by different sectors in this study could be based upon different underlying values held by leading stakeholders. Research has already indicated that 'community energy' initiatives are variously driven by instrumental, substantial and normative goals (Walker, Hunter, Devine-Wright, Evans, & Fay, 2007) sometimes existing to redress local problems, other times used instrumentally to encourage social acceptance of large-scale infrastructure projects. Similarly, urban DE is plural and diverse, coming in different guises from different sectors. Given these, it is important for policy-makers to recognise this diversity when devising new regulations, planning policies or financial mechanisms to encourage further growth of this sector.

While some drivers confirm the notion of DE as a predominantly local practice (e.g. helping the poor, improving the housing stock), other drivers were clearly of a less local nature (e.g. helping the environment, researching and demonstrating). This suggests that local motivations are not exclusively relevant to community sector projects, but shape projects across the entire DE sector while co-existing alongside non-local drivers. The case studies confirmed previous findings about the importance of funding as an influence upon DE initiatives (Adams & Berry, 2008; Allen, Hammond, & McManus, 2008; Julian & Dobson, 2012), and also reveal that this is a sector-specific finding – private sector projects were less affected. Both private sector-led cases did not represent funding as a problematic issue; they were funded as investments, not through grants, and thus were not inhibited by the funding-related difficulties (restrictive grant stipulations, grant-related deadlines, being relatively unprepared due to sudden funding availability) mentioned by the community, public and professional third sectors.

Although lack of trust did not hinder these particular projects in the sense that none of them ran into strong opposition fuelled by mistrust, only projects in the community and third sector were represented as specifically benefitting from local residents' trust. Therefore, the previous conclusions about trust as an important but not universally present characteristic of energy projects labelled 'community' (Walker, Devine-Wright, Hunter, High, & Evans, 2010) seem to ring true perhaps more for community and third sector projects than for public and private sector DE initiatives. Stakeholders from the four sectors differed in terms of attempts to engage local energy users, which was less common outside of the community and third sector. Stakeholders across the four sectors perceived local publics to lack awareness, understanding and interest in energy issues, a pattern reported before (Barnett et al., 2012); however, while in the public and private sector this led to a lack of engagement altogether, community and third sector stakeholders indicated much more willingness to attempt to work towards 'energy citizenship' (Devine-Wright, 2007).

In terms of policy implications, the plurality of ways in which DE is applied and its variety in underlying drivers and outcomes suggest that its potential should not be evaluated merely through a carbon or energy production lens, but instead by looking across environmental, social and financial dimensions. Moreover, these findings highlight the heterogeneity of the DE sector and reveal some of the ways in which community energy is very different from DE led by stakeholders from the public, private or professional third sectors. This suggests the need for a critical viewpoint regarding the relative merits of encouraging community-led, rather than, for example, public sector-led DE. While community-led DE may benefit from higher levels of trust due to being more 'local' (Devine-Wright & Wiersma, 2013) and not for profit, as well as possibly enhancing public awareness of energy issues, it also reinforces dependence on grant funding and reliance upon local leaders or project champions (Hoffman & High-Pippert, 2010), which may not be viable in all areas due to differences in local capacities. An overt focus upon community DE overlooks private sector independence from government grants. While the diversity of drivers behind the emergence of these case study projects suggests that multiple policy goals can be achieved simultaneously, this multiplicity also means that some policy goals resonate more with the way some sectors conceive DE. Private sector-led initiatives are more likely to achieve policy goals for DE that focus upon increasing deployment of renewable energy technologies; community, public and third sector initiatives are more likely to address policy objectives for housing quality, urban regeneration, research and development, fuel poverty, social cohesion and behavioural change.

This study's adoption of a comparative case study design has been successful in opening up the differences between DE led by four different sectors; however, one limitation is that its conclusions are inevitably based on a limited number of cases and a limited number of in-depth interviews per case. Furthermore, this study, as well as others (Adams & Berry, 2008; Allen et al., 2012; Walker et al., 2007), studied projects that were relatively successful (i.e. they managed to come to fruition despite a generally unsupportive context). Understanding the dynamics of projects that were less successful (i.e. either by failing to develop altogether or by struggling to achieve the desired results) would provide an important contrasting perspective to achieve a fuller understanding of the dynamics of more DE.

In summary, the findings suggest that DE encompasses projects driven and influenced by a range of different factors. Drivers of DE were highly diverse, varying by sector and were predominantly local, with addressing poverty predominant. Key influencers upon the development of DE projects included the availability of funding (and conditions attached), levels of (mis)trust and stakeholder perceptions of the capacities and knowledge of energy users. In conclusion, policy and academic emphasis on community-led DE is too narrow, overlooking other successful configurations that can contribute to the low carbon transition that is integral to successful climate change mitigation.

Acknowledgements

We thank all participants our project colleagues and three anonymous reviewers for their helpful comments.

Funding

This study was undertaken as part of the CLUES project (Challenging Lock-in to Urban Energy Systems) supported by funding from the Engineering and Physical Sciences Research Council under the Sustainable Urban Environments programme (Grant ref.: EP/I002170/1).

Notes

1. In this paper, we focus on 'meso-level' DE, which sits between the micro-level (i.e. individual or household-level microgeneration) and the macro-level (i.e. large-scale centralised generation) (Walker & Cass, 2007).
2. For one of the nine case study projects, this research drew on interview transcripts available through a separate piece of research, which was suitable due to the similarity in interview protocol and methods.
3. Names have been anonymised throughout this paper.

References

Adams, S., & Berry, S. (2008). *Low carbon communities: A study of community energy projects in the UK*. Ruralnet: Warwickshire.

Aiken, G. (2012). Community transitions to low carbon futures in the transition towns network (TTN). *Geography Compass, 6*(2), 89–99.

Alanne, K., & Saari, A. (2005). Distributed energy generation and sustainable development. *Renewable and Sustainable Energy Reviews, 10*, 539–558.

Allen, J., Sheate, W. R., & Diaz-Chavez, R. (2012). Community-based renewable energy in the Lake District National Park – local drivers, enablers, barriers and solutions. *Local Environment, 17*, 261–280.

Allen, S. R., Hammond, G. P., & McManus, M. C. (2008). Prospects for and barriers to domestic microgeneration: A United Kingdom perspective. *Applied Energy, 85*, 528–544.

Arentsen, M., & Bellekom, S. (2014). Power to the people: Local energy initiatives as seedbeds of innovation? *Energy, Sustainability & Society, 4*(2). Retrieved from http://www.energsustainsoc.com/content/4/1/2

Barnett, J., Burningham, K., Walker, G., & Cass, N. (2012). Imagined publics and engagement around renewable energy technologies in the UK. *Public Understanding of Science, 21*(1), 36–50.

Bergman, N., & Eyre, N. (2011). What role for microgeneration in a shift to a low carbon domestic energy sector in the UK? *Energy Efficiency, 4*, 335–353.

Bomberg, E., & McEwen, N. (2012). Mobilizing community energy. *Energy Policy, 51*, 435–444.

Bouffard, F., & Kirschen, D. S. (2008). Centralised and distributed energy systems. *Energy Policy, 36*, 4504–4508.

Braun, V., & Clarke, V. (2006). Using thematic analysis in psychology. *Qualitative Research in Psychology, 3*, 77–101.

Butler, C., Parkhill, K. A., & Pidgeon, N. (2013). *Deliberating energy transitions in the UK - Transforming the UK energy system: Public values, attitudes and acceptability*. London: UKERC.

Catney, P., MacGregor, S., Dobson, A., Hall, S. M., Royston, S., Robinson, Z., ... Ross, S. (2014). Big society, little justice? Community renewable energy and the politics of localism. *Local Environment, 19*, 715–730.

Department for Business, Enterprise and Regulatory Reform. (2008). *UK renewable energy strategy – consultation*. London: The Stationery Office.

Department of Energy and Climate Change. (2011). *Microgeneration strategy 2011*. London: DECC.

DECC. (2014). 2013 UK greenhouse gas emissions, provisional figures.

Devine-Wright, P. (2007). Energy citizenship: Psychological aspects of evolution in sustainable energy technologies. In J. Murphy (Ed.), *Governing technology for sustainability* (pp. 63–86). London: Earthscan.

Devine-Wright, P., Walker, G., Hunter, S., High, H., & Evans, B. (2007). *An empirical study of public beliefs about community renewable energy projects in England and Wales* (Community Energy Initiatives Project Working Paper 2). Retrieved from http://geography.lancs.ac.uk/cei/CommunityEnergyKeyPubli cations.htm

Devine-Wright, P., & Wiersma, B. (2013). Opening up the 'local' to analysis: Exploring the spatiality of UK urban decentralised energy initiatives. *Local Environment, 18*, 1099–1116.

Greenpeace. (2005). *Decentralising power: An energy revolution for the 21st century*. London: Greenpeace.

Hargreaves, T., Hielscher, S., Seyfang, G., & Smith, A. (2013). Grassroots innovations in community energy: The role of intermediaries in niche development. *Global Environmental Change, 23*, 868–880.

Hathway, K. (2010). *Community power empowers*. London: Urban Forum.

Hoffman, S. M., & High-Pippert, A. (2010). From private lives to collective action: Recruitment and participation incentives for a community energy program. *Energy Policy, 38*, 7567–7574.

Intergovernmental Panel on Climate Change. (2014). *Fifth assessment report: Working group III – mitigation*. Chapter 7 Energy Systems.

Joffe, H. (2011). Thematic analysis. In D. Harper & A. R. Thompson (Eds.), *Qualitative research methods in mental health and psychotherapy: A guide for students and practitioners* (pp. 209–223). Chichester: John Wiley & Sons.

Julian, C., & Dobson, J. (2012). *Re-energising our communities*. London: ResPublica.

Mulugetta, D., Jackson, T., & Van der Horst, D. (2010). Carbon reduction at community scale. *Energy Policy, 38*, 7541–7545.

Sauter, R., & Watson, J. (2007). Strategies for the deployment of micro-generation: Implications for social acceptance. *Energy Policy, 35*, 2770–2779.

Seyfang, G., Park, J. J., & Smith, A. (2013). A thousand flowers blooming? An examination of community energy in the UK. *Energy Policy, 61*, 977–989.

Turcu, C., & Rydin, Y. (2012). Planning for change in urban energy systems. *Town & Country Planning, 81*, 272–232.

Walker, G. (2008). What are the barriers and incentives for community-owned means of energy production and use? *Energy Policy, 36*, 4401–4405.

Walker, G., & Cass, N. (2007). Carbon reduction, 'the public' and renewable energy: Engaging with socio-technical configurations. *Area, 39*, 458–469.

Walker, G., Devine-Wright, P., Hunter, S., High, H., & Evans, B. (2010). Trust and community: Exploring the meanings, contexts and dynamics of community renewable energy. *Energy Policy, 38*, 2655–2663.

Walker, G., Hunter, S., Devine-Wright, P., Evans, B., & Fay, H. (2007). Harnessing community energies: Explaining and evaluating community-based localism in renewable energy policy in the UK. *Global Environmental Politics, 7*, 64–82.

Warren, C., & McFadyen, M. (2010). Does community ownership affect public attitudes to wind energy? A case study from south-west Scotland. *Land Use Policy, 27*, 204–213.

Watson, J., & Devine-Wright, P. (2011). Centralization, decentralization and the scales in between. In M. Pollitt & T. Jamasb (Eds.), *The future of electricity demand: Customers, citizens and loads* (pp. 542–577). Cambridge: Cambridge University Press.

Urban experiments and climate change: securing zero carbon development in Bangalore

Harriet Bulkeley[a] and Vanesa Castán Broto[b]

[a]*Department of Geography, Durham University, Durham, UK;* [b]*Development Planning Unit, UCL, London, UK*

Climate change is an increasingly important issue on urban policy and research agendas. As this agenda gathers pace, this paper argues for an approach that recognises the critical role of climate change experiments in meditating the response to climate change in the city. Drawing on a case study of a green housing development in the outskirts of Bangalore in India—Towards Zero Carbon Development (T-Zed)—the paper follows the emergence of an experiment in the simultaneous processes of making, maintaining and living low carbon alongside and in between existing infrastructure regimes. It is argued that this experiment has created space for social and technical innovation, reworking notions of urban development in Bangalore. At the same time, it has reconfigured existing urban infrastructure networks through new discourses and practices of urban ecological security, enabling the emergence of a new rhetoric of low carbon living within the city that effectively marries green forms of consumption with urban development. While the experiment serves as a means for modifying urbanism in Bangalore, its results are ambivalent in the context of ongoing inequalities within the city and beyond.

Introduction

For many, climate change is a problem of global proportion, requiring equally global responses. Yet climate change is also a fundamental urban issue. With an increasingly urban population, cities are places which may be particularly vulnerable to the impacts of climate change and—the focus of this paper—are significant sources of greenhouse gas (GHG) emissions. Indeed, cities and towns are thought to produce over 70% of global energy-related carbon dioxide emissions, and by 2030 some 80% of the increase in global annual energy demand above 2006 levels is predicted to be from cities in non-Organisation for Economic Co-operation and Development (OECD) countries (IEA, 2008). Reflecting the twin challenges of mitigating and adapting to climate change in the city, since the mid-1990s scholars have documented the emergence and institutionalisation of climate change as an issue of and for urban governance (Betsill & Bulkeley, 2007; Bulkeley, 2010; Hodson & Marvin, 2009; Monstadt, 2009; Schreurs, 2008; While *et al.*, 2010). For the most part, work in this field has been concerned with assessing the development and implementation of urban climate policy. Central to these analyses has been the investigation of the policies and measures that are being developed in key urban infrastructure networks: energy, transport, the built environment, and increasingly water and sanitation. However, infrastructure networks, their material fabric, as well as the everyday practices and political economies that sustain and are structured by them have remained outside the scope of analysis. This is a critical omission since these networks

> structure a major part of the material metabolism in industrialized societies. They source, use, and transform huge amounts of natural resources. At the same time they are key catalysts of environmental problems like air, water, and soil pollution, and nuclear risks, and they make a major contribution to global warming. (Monstadt, 2009, p. 3)

The present paper seeks to develop an alternative account of urban responses to climate change, one that makes space for considering the role of infrastructure networks in mediating governance (Bulkeley *et al.*, 2011; Graham & Marvin, 2001; Hodson & Marvin, 2009; Monstadt, 2009). Rather than being orchestrated solely through traditional forms of governing—plan-making, regulation and implementation—it is argued that governing climate change in the city also involves alternative interventions—including, for example, partnerships, demonstration and pilot projects, and area-based initiatives—which are collectively termed *climate change experiments* in urban socio-technical systems. Research on climate change and cities has identified an ever-increasing range of initiatives in the name of climate change taken by both public and private actors. Here, the paper moves from examining the factors that have generated climate action to examining the form of such action and its impact on the fabric and politics of the city. Drawing on a case study of a green housing development in the outskirts of Bangalore in India—Towards Zero Carbon Development (T-Zed)—the paper follows the emergence of a low carbon urban infrastructure mode, alongside and in-between existing infrastructure regimes, in simultaneous processes of making, maintaining and living low carbon. The rest of the paper first introduces in more detail the notion of climate change experiments, before turning in detail to the case of T-Zed. The aim is to understand the ways in which such experiments govern climate change in the city, and their potential to catalyse a systemic and just transition towards low carbon urbanism, while recognising that such transitions are only part of a much broader agenda for sustainable development.

Making space for climate change experiments

Initially focused on issues of climate mitigation and led by municipal networks based in Europe and North America, the urban climate agenda has recently been extended and invigorated by the growing involvement of 'global' and mega-cities North and

South, increasing attention to issues of adaptation, and a more overtly political discourse concerning the capacity of cities to lead where national and international processes to respond to climate change appear to be faltering (Bulkeley, 2010; Satterthwaite, 2011). For the most part, academic analyses of this phenomenon have focused on the emergence of climate change as an issue of urban policy, the multilevel governance contexts and transnational networks that have given rise to the urban climate change agenda, and the institutional and political factors that have shaped the (growing) gap between the rhetoric of city actors and climate action (for example, Bai, 2007; Bulkeley & Kern, 2006; Gore & Robinson, 2009; Granberg & Elander, 2007; Holgate, 2007; Puppim de Oliveira, 2009; Romero Lankao, 2007; Sugiyama & Takeuchi, 2008). As scholars have illustrated, underpinning the increasingly widespread uptake of climate change at the urban level has been a commitment to a classical evidence-based approach to policy-making. In research and policy communities alike, the argument is made that the first stage of any urban response must be to understand the risks of climate change, either in terms of urban models of GHG emissions or potential climatic hazards, before setting targets, devising plans, and implementing action (Corfee-Morlot *et al.*, 2009; Dhakal, 2009). Despite the intuitive appeal of this linear narrative, the reality of urban climate change responses for the most part fails to live up to such expectations. While many cities have sought to model their current and future emissions of GHG and have set targets, comprehensive action plans which respond to this evidence are few and far between. Analysing this situation, Alber & Kern (2008, p. 4) conclude that 'numerous cities, which have adopted GHG reduction targets, have failed to pursue such a systematic and structured approach and, instead, prefer to implement no-regret measures on a case by case basis' (also Jollands, 2008). Viewed through the lens of a policy model in which evidence, goals, planning and action flow seamlessly, the failure of cities to translate knowledge into policy, and the gaps between policy rhetoric and implementation, call for an assessment of the barriers encountered, often found in terms of a lack of institutional capacity (knowledge, know-how and resource, for example). Here, the ad-hoc and project-based responses which are emerging in cities seemingly go either unnoticed or are regarded as a mark of the failure of cities really to engage with the climate change agenda.

Whilst not wanting to undermine the importance of understanding the factors that are shaping the capacity for urban responses to climate change, this paper suggests that remaining wedded to notions of policy-making that privilege the relation between knowledge, institutions and resources above issues of political economy and the social and technical dynamics of cities is providing an impoverished picture of the challenges facing urban climate governance. Recognising governance as a more contested, partial and fragmented process (Rutland & Aylett, 2008), suggesting that one needs to look 'off-plan' for the sites in which governing climate change is taking place. In his recent analysis of the dynamics of climate governance, Hoffman (2011) argues that we have entered an era of 'governance experimentation', driven by the twin pressures of disillusionment with the process of mega-multilateralism, though which international agreements on climate change have conventionally been reached, and the fragmentation of political authority across a range of public and private actors. Governance experiments, he suggests, are 'rule-making endeavors in non-traditional political spaces' (Hoffman 2009, p. 4), or what Hajer (2003) terms policy-making in the absence of polity. While Hoffman's analysis explicitly excludes experiments within individual cities, examining the ways in which climate governance is taking place in the city within and between public/private actors suggests that there may be many instances where governing climate change is taking place in 'non-traditional political spaces'. Such interventions are experimental in so far as they are 'innovative' and imply 'trial and error' with novel governance mechanisms (Hoffman, 2011), but also because of their often tentative nature, the sense of 'testing' or establishing (best) practice that frequently accompanies their development, and the ways in which they are used as a means of supporting knowledge claims and discursive positions.[1] Rather than regarding ad-hoc arrangements, individual schemes, exemplars or best-practice projects as accidental to the main game of governing climate change in the city, this analysis suggests that such interventions may be symptomatic of forms of climate governance.

One striking feature about such climate change experiments at the urban level, perhaps in contrast to the other arenas within which Hoffman's analysis has been undertaken, is the ways in which they are orchestrated around infrastructure networks. Although analyses of urban responses to climate change have drawn attention to the ways in which GHG emissions are produced through infrastructure networks (including energy, water, waste, sanitation and transport, for example), the ways in which such socio-technical systems mediate potential governance responses has frequently been neglected. Within the body of literature that considers socio-technical systems and their transformation, researchers have adopted a 'multilevel' perspective for understanding systems dynamics, comprising the landscape, which 'provides the macro-level structuring context', socio-technical regimes which 'constitute the mainstream, and highly institutionalised, way of currently realising societal functions', and niches, relatively protected spaces within which innovation takes place (Smith *et al.*, 2010, p. 440). Here, too, forms of experimentation appear to be critical sites within which shifts in existing socio-technical networks can be achieved, for

> [h]istorical experience suggests radical changes begin within networks of pioneering organizations, technologies and users that form a *niche* practice on the margins of the regime. Studies suggest these 'niche' situations (e.g. niche applications, demonstration programmes, social movements) provide space for new ideas, artefacts and practices to develop without being exposed to the full range of selection pressures that favour the regime. (Smith, 2007, p. 429)

In drawing attention to the potential role that 'niche' or experimental interventions might play in reshaping urban infrastructure regimes, this literature also points to the potential value in looking at the forms of urban climate governance that are taking place 'off plan'.

Working between the notions of 'governance experiments', on the one hand, and 'niche' innovation, on the other, provides an alternative perspective on the potential significance of the growing number of ad-hoc interventions emerging in cities in response to climate change. Critically, this paper suggests that such experiments need to be viewed as socio-technical, that is as co-produced through the interrelation of their social and material elements, and as providing a means through which the governing of climate change is conducted. As other authors have pointed out, urban infrastructure networks are a critical means through which political, economic, and ecological structures and practices are reproduced and reconfigured (Keil, 2005; Monstadt, 2009). In seeking to affect a low carbon mode of urbanism, climate change experiments may serve to reproduce or realign such 'political ecologies'. While *et al.* (2010, p. 82) suggest that processes of eco-state restructuring are now focused on 'carbon control', creating a 'distinctive political economy associated with climate mitigation in which discourses of climate change both open up, and necessitate an extension of, state intervention in the spheres of production and consumption.' As this politics of carbon control 'comes to ground' the 'the calculative practices of urban management' are shifted and new forms of financial strategy and economic development enacted, a process in which 'experiments in the reterritorialisation of governance at the city-regional scale around carbon interdependencies linked to energy supply, infrastructure, or the pooling of carbon credits' may be significant (While *et al.*, 2010, p. 87). In a similar vein, Hodson & Marvin (2009, pp. 195–196) suggest that issues of climate change mitigation and adaptation are becoming a key strategic concern for urban authorities, provoked by discourses of the urban causes and consequences of environmental problems and facilitated through the restructuring of the state and the creation of 'new state spaces'. This, they contend, is leading 'the world's largest cities' to begin 'to translate their strategic concern about their ability to guarantee resources into strategies designed to reshape the city and their relations with resources and other spaces' (p. 200). Rather than regarding experiments as 'niches', somehow outside of existing regimes, these analyses point to their critical and ambivalent role in both sustaining existing urban political ecologies and offering the potential for alternative urban futures.

Far from representing the poor relation to an ideal urban climate policy, a reading of literatures on climate governance, socio-technical systems and urban political ecologies suggests that 'experiments', albeit differently theorised, offer an important means through which to understand urban climate governance. In order to understand these dynamics, it is proposed that the meaning and potential of climate change experiments can be understood through three, related processes: making, maintaining and living low carbon. 'Making' relates to the process of assembling material and semiotic networks to accomplish an experiment, both in terms of creating new arrangements and building legitimacy for alternatives outside mainstream channels of policy (cf. Hoffman, 2011; Hajer, 2003) and in terms of demonstrating how it operates in practice, in a similar way to which niches operate in the multilevel perspective (Geels, 2002). 'Maintaining', in contrast, refers to the processes of readjustment that take place in order to deal with the experiment within the political economy and political ecology of the city, in terms of both realigning low carbon interventions with existing urbanisation priorities (Rutland & Aylett, 2008) and reconfiguring systems of provision to adapt to new demands (Monstadt, 2009). 'Living' refers to the processes whereby the experiment is integrated into everyday practices, particularly in relation to the expectations of those citizens whose lives are intertwined with the daily operation of the experiment, but also in relation to the ways in which the intervention becomes part of the normal operating practices of its stakeholders. This involves new forms of conduct and normalisation, which may conflict with existing subjectivities and through which alternative notions of what the experiment is and what it should become may come into conflict. The next section introduces the case of Towards Zero carbon Development (T-Zed) in Bangalore, before considering these processes in turn.

Towards Zero-Carbon Development (T-Zed) in Bangalore: making, maintaining and living low carbon

For the most part, considerations of climate change and cities in the global South have focused on the pressing challenges of climate vulnerability and the way in which it is coproduced through existing political economies and persistent inequalities (Romero Lankao, 2010). Commentators have pointed to the inadequacy of existing infrastructure provision and forms of governance for addressing current vulnerability and the ways in which these 'deficits' are profoundly limiting capacity for climate adaptation (Satterthwaite, 2011). In the Indian context, Revi (2008, p. 211) argues that 'a chasm exists between the official urban "city building" development agenda and vulnerability reduction for those most at risk in these urban areas' such that 'the current scale of demolitions and relocations is compounding the vulnerability of many urban residents'. Given this context, and the imperative of responses to climate change that take account of those most likely to be affected by it (World Bank, 2010), concerns have been raised that discourses which attribute responsibility for GHG emissions to cities may both over-exaggerate urban contributions and smooth over the significant differences between cities (Dodman, 2009; Satterthwaite, 2008). At the same time, recognising differences between cities in terms of their contribution to GHG emissions also entails acknowledging the differences within cities in relation to per capita emissions (Hoornweg et al., 2011), and consequently that it may be appropriate to explore some forms of low carbon urban development as one part of pursuing sustainable development in cities in the global South.

It is in this context of complex urban landscapes of climate change vulnerability and responsibility that the case study of Bangalore is situated.[2] Bangalore is a rapidly growing city that has undergone a profound social and economic transformation in the last two decades, associated with the rise of the information technology (IT) industry. This has had a strong impact on the urban fabric, for example by developing new communication infrastructures and the building of new developments on the edge of the city to serve the needs of a rapidly emerging industry and middle-class population. These new gated communities are putting pressure on the city's resources, in relation to the supply of energy, but particularly with respect to water. The limited municipal water supply system only extends to planned areas of development, and new private developments have to deal with the lack of infrastructure and the need to pay water at higher prices in the private markets, or, where they can afford it, opening boreholes to tap into the water table, reducing water availability in the long-term for the city's wider population (Ranganathan *et al.*, 2009). At the same time, while slums are often regarded as a minor problem compared with other cities in India, activist groups working in slums, however, differ. For example, according to the Alternative Law Forum in 2005 there were 778 slums in Bangalore housing 26% of the population, and 35% of the population in Bangalore could be considered as urban poor (CIVIC, 2008).

In this context of rapid urbanisation and resource constraints, Towards Zero Carbon Development (T-Zed) is a project that seeks to provide an alternative for higher-income classes in Bangalore to tap into 'responsible housing' and create an example that can be replicated in the city. The development is managed by a private developer, Biodiversity Conservation India Limited (BCIL), committed to green values. While the development takes the form of a gated community on the outskirts of the city, a familiar type of urban development in the city, it incorporates numerous social and technical innovations with the intention of reducing the carbon emissions of the development and its dependence on the city's resources (BCIL, 2009). The design included considerations to reduce the dependence of the development from the city's water and electricity supply, measures to reduce the embodied energy of the building (including the use of local materials), and emphasis on landscaping and respecting the existing vegetation in the original plot. This constitutes a clear alternative to mainstream developments in Bangalore. First, in terms of use of materials, conventional housing relies on the use of materials with high embodied carbon such as marble and glass, which are produced energy-intensive processes and usually transported from large distances. Second, mainstream design in Bangalore has little regard for considerations of efficient use of energy and water or waste recycling. Yet, in T-Zed there is an attempt to enable residents to maintain lifestyles like those associated with mainstream developments by providing them spacious and light apartments and even a swimming pool. The development includes 16 single-family houses and 75 apartments and in March 2010 it was almost fully occupied. As articulated in further detail below, through both acting as a new political site within which discourses of what low carbon development within the Bangalore context might entail and through the process of assembling a new constellation of social and technical relations, T-Zed can be considered as a climate change experiment.

The account of T-Zed developed in this paper incorporates the findings of a review of relevant policy documents and grey literature and 36 interviews with individuals and groups carried out in Bangalore in March 2010. The interviews were conducted in English, using an interview guide which covered existing climate change initiatives in the city and included concrete questions about T-Zed. The interviews lasted between 30 minutes and 2 hours, and in several cases included a visit to the location of the initiative and a demonstration of the technologies involved. Of the interviews, 19 were specifically about T-Zed, being conducted with members of BCIL and other low carbon housing developers (D = 6), T-Zed residents (R = 5) and knowledge providers and consultants who had, directly or indirectly taken part in T-Zed (C = 8). Additional interviews were carried out with government officials of the government of Karnataka (G = 2); non-governmental organisations that work in Bangalore (N = 6); think tanks and universities (T = 6), and international organisations operating in cities in India, such as the Climate Group (I = 3). Most of these interviewees were not aware of T-Zed, but they provided valuable information to understand the context of low carbon governance and urbanism in Bangalore and India. A thematic analysis was developed after coding the data in Nvivo, with the aim of understanding the processes through which climate change experiments like T-Zed are formed and their wider implications for the emergence of a low carbon mode of urbanism in Bangalore. Following insights derived from the literatures on urban climate governance, socio-technical systems and urban political ecologies, as described above, the analysis was then developed to examine the instances in which T-Zed may be contributing to making, maintaining and living low carbon in Bangalore. These results are discussed below, using material from the interviews as a means of illustrating the discursive positions taken by different actors about the T-Zed development and its role within the wider processes and politics of urbanisation in Bangalore.

Making low carbon

In the case of T-Zed, two parallel developments intervened in creating a space for experimentation: the formation of a suitable discourse of low carbon innovation within the urban governance context of Bangalore, and the construction of a suitable constellation of actors who supported and replicated those discourses.

BCIL was founded by Chandrasekar Hariharan, an economist who worked in the charity sector until the early 1990s. Together with his management team, Hariharan present himself as generating the discourse of low carbon innovation in housing within Bangalore. As BCIL declares on its website, 'the only Boss they recognize is the Big Idea', i.e. BCIL's objective both as a 'vision' and 'duty' was to mainstream sustainability into the conventional building sector. However, 'the idea' alone will not open up spaces of experimentation if it is not closely fitted within the urban governance context. Rather than seeking to tackle the widespread inequalities evident in the city, BCIL focused its interpretation of urban sustainability on demonstrating the potential of 'green' gated communities as a means of tapping into the aspirations of the growing professional classes in the city:

> There are people who are in various [professional] sectors. They appreciate people like us doing [sustainable buildings] ... they want to be a part of those kinds of developments. So we make an opportunity for them to come down and be part of it ... be members of this particular project, be owners of this particular project. So, they buy into these campuses. [...] We build homes for each individual, trying [to] give them that experience of 'Green homes.
> (Interview D1)

In the process of making its green building concept, BCIL started building 'second homes' in low-density developments in the mid-1990s, in what can be described as an incubation period. This provided the opportunity to refine both the product that

BCIL was selling and its core values of leading design and innovation, towards the launch of the T-Zed project in 2002. In effect, T-Zed was the first attempt from BCIL to come out of the 'lab' of individual houses and expand into a larger scale. The challenge, as BCIL managers saw it, was to develop a model of green development that could compete with regular urban homes, thus moving away from building individual flagship projects. BCIL purchased a plot of farm land in the area of White-field, in the east edge of Vartur Road. The location was ideal for developing a project that would make green housing available for middle- to upper-class families working in the IT sector, in a growing, thriving area at the edge of the city. In this context, as a former BCIL worker explains, they proceeded by

> calculating the normal size of homes which would sell like hot cakes [...] at this kind of budget then say about 1000–1500 square feet. This is usually a safe bet because a predo-minant bulk of professional Bangalore is middle class and most of them in that middle class segment aspire for more but are comfortable [with 1000 square feet] ... and that's the bulk of today's buying population or people investing in the real estate sector. (Interview C4)

The significant social and economic transformation that has accompanied the growth of the IT sector in Bangalore has not only increased the gross domestic product (GDP) per capita of the population, but also signalled the emergence of a distinct and cosmopolitan middle-class, young and educated, associated with the IT industry. These changes have led to a new understanding of what the city is and what it should provide. It is in this context that the 'Green Gated Community' offers 'a bridge between living and making a living' (Interview D3): idyllic landscapes, secure setting and access to the city's economic centres. Indeed, BCIL sold initially 16 'Candida' single-family houses which financed the development of the apartments that comprise the rest of the development. Overall the houses sold 'because the location is right, the finish is right, the cost is right' (Interview D3).

In seeking to make a constituency for low carbon urbanism in Bangalore, developers explicitly linked 'the idea' of sustainable homes with the need for the middle classes to take responsibility for climate change. For example, a former BCIL worker explained that in projects like T-Zed they were 'trying to create small examples out there and [...] tell people: "okay; you guys have enough money ... be a part of a [sustainable] project"' (Interview D1). Residents in T-Zed were described as

> eco-friendly people ... who have it in their heart that they need to do something for this community ... people from all backgrounds, the cream of the society: from software [professionals] to doctors to people in the retiring ages .... (Interview D2)

'The idea' of urban sustainability is here invoked as a way to solve the contradiction between emerging middle-class lifestyles and the responsibilities for addressing climate change. Residents also emphasised that they were attracted by the low carbon concept put forward in T-Zed. Critical to making low carbon development was the notion that sustainability could be aspirational. There is no perceived contra-diction between big homes and sustainable homes, and indeed, as residents highlighted:

this enables us to live the lifestyle we want you, we do not compromise on our comforts at all, but at the same time it is important to think about the quality of life and living within this community, being responsible for climate change. (Interview R3)

Making T-Zed as a climate change experiment required BCIL to draw a constellation of actors around 'the idea', including a set of new technologies and customers who had a particular interest in these forms of innovation. Enrolling customers in an expensive low carbon development requires taking advantage of the existence of a cosmopolitan upper-middle class who are not only concerned about low carbon, but also have the economic resources to pay for it. Hence, the process of making this experiment is predicated on the opportunities open by existing urban inequalities—the existence of a clear divide between the localised economies of the urban poor and the cosmopolitan economies of the higher-middle classes—and the urban governance structures which reproduce these inequalities. BCIL present themselves as providing an alternative to conventional development as 'good' or 'sustainable', without challenging existing consumption patterns in housing or addressing their social consequences including the splintering of urban systems (Graham & Marvin, 2001) and loss of livelihoods for the poor in peri-urban areas through changing patterns of land ownership (Keivani & Mattingly, 2007).

Another set of actors (designers, architects, engineers, entrepreneurs) simultaneously mediates the translation of 'the idea' into a variety of socio-technical innovations that aim at easing the tension between lifestyles and environmental responsibility in different ways. Here, a particularly important feature driven by Bangalore's recent urbanisation has been the lack of control of the local and state governments over the city's development, particularly at its periphery, which gives 'the flexibility to experiment as much as possible and, because you have a commitment to the client and the market, you can't have a failure' (Interview C3).[3] Sometimes this occurred by following ideas of other innovators within the city. For example, one of the main features of T-Zed is the use of compressed-earth blocks and other natural materials, a technique pioneered by the architect Chitra Vishwanath in Bangalore since the end of the 1980s, drawing inspiration from the organic architecture concepts for low-cost and low-energy buildings developed by Laurie Baker. These traditional concepts alone, however, are insufficient to meet the energy and resource demand of BCIL customers in aspects such as thermal control or appearance. Thus, T-Zed also provided an experimentation space for new actors who were trying to develop high-cost, low-carbon technologies at a commercial scale, and which functioned to draw these disparate actors into a common purpose. One of the founders of BCIL explains: 'innovation is only marginal! ... You know what we do? We are very smart technology scouts! We are constantly scouting for the right technologists!' (Interview D3). For example, one of the contractors in T-Zed, Flexitron, developed several light-emitting diode (LED) lighting options for T-Zed. When T-Zed started, LED 'was at its rudimentary stages and I remember us going through some disastrous design options' (Interview C4). A suitable option was found and today Flexitron is a successful company. As its Director explains:

> T-Zed was ... what to say ... responsible to trust us to offer them a technology for something very new! And, it is like me buying a car from a total new company! I don't know what brand of car it is or how it works or nothing ... it has no background ... and I say no, here's the money, you still supply me! So that's what T-Zed did! ... with this, T-Zed supported us to mature this technology! (Interview C5)

Making T-Zed involved successfully aligning sustainable housing ideas with the demands of an emerging middle class, serving to forge a collective of customers who would buy into the project and accomplished through exploiting the 'absences' in the urban governance context to create opportunities for social and technical innovation. While there are undoubtedly differences between actors in terms of how these issues are framed, it is through this constellation of actors and technical innovation that 'the idea' of low carbon urbanism, now enacted in T-Zed, came to be produced and may be replicated at a wider scale. For leading actors in the city, 'T-Zed has set a trend in Bangalore' (Interview C7) and the Green Building niche in Bangalore appears to be growing as 'there are a lot of builders who got influenced because of T-Zed, because of BCIL' (Interview C6). In this manner, T-Zed has succeeded in making both the idea of low carbon housing and its practical application part of Bangalore's future urban development.

At the same time, there are actors in the broader context of urban development who do not subscribe to this common discourse of the ecological motivations behind T-Zed. Such actors often articulate alternative conceptions of sustainability, beyond achieving low carbon or resource sufficiency, and in particular the need to achieve a just city, as part of a sustainable one, was highlighted on numerous occasions. For example, a representative of a local non-governmental organisation (NGO) highlighted that sustainability 'is about sharing the resources fairly ... and in view of the fact that we need to conserve them and preserve them for future generations' (Interview N3). Rather than regarding low carbon urbanism as part of mainstream housing development, for these actors it held the potential for enhancing the access of the poor to energy resources. For example, the Chief Executive Officer (CEO) of Flexitron explained that experimentation in T-Zed had provided the basis for low-consumption solar lamps for people with limited or no access to electricity services. Yet, concerns for the livelihoods of the poor remain excluded from the dominant discourse of low carbon urbanism: as the CEO of BCIL stressed, 'I am not ... in the sphere of social change, which is: can I make these places affordable for the poor? ... that's not my business!'

Maintaining low carbon

Making low carbon alone is not sufficient: ideas and actor constellations can create the space for experimentation, but for the experiment to take hold a series of sociotechnical relationships need to be created and renegotiated. In this case, we consider the process as one of *maintaining* the low carbon nature of the experiment, both in terms of T-Zed as a 'green' gated community, and the expectations of residents and technologies that requires, as well as in terms of the relation between T-Zed and the urban networks within which it is situated. In these processes, the development of the experiment and its insertion within the city involves a series of splintering and rebundling processes through which urban infrastructures are reconfigured, a process which resonates strongly with the splintering urbanism thesis, which links the rise in urban inequalities resulting from the fragmentation of the urban infrastructure and its concentration in privileged spaces with the institutional and economic changes resulting from globalisation and the transnational movement of capital (Graham & Marvin, 2001).

Maintaining T-Zed as a 'green' gated community has entailed ongoing work. Although it appears that originally BCIL intended to deliver a completed project, today they argue that BCIL needs to take care of the maintenance of the project by providing customised care. Setbacks in the building process, at least in part the result of the experimental use of materials and technologies, led to a two-year delay in the delivery of the project and work in 'finishing' the project is ongoing, creating a direct role for BCIL in the day-to-day management of the site. Perhaps more fundamentally, managing the material and technical systems that operate in the compound has required progressive adjustments to fit technologies to local contexts. Partly, this is because some of the technologies were customised to such a level that only BCIL-trained personnel can manage them. In other cases, technologies have simply not worked, as it is the case of the custom-made refrigerators and the air-conditioning system. As a result, managing the development has involved looking for additional solutions, and thus the process of design and innovation is continually unfolding. It seems that the experimental nature of the project has created a need for constant maintenance, in turn changing the relation between developers and residents from one of producer and consumer to a, sometimes uneasy and contested, continuing partnership. As well as working to care for the technical side of the development, BCIL have also sought to manage the expectations of residents and adjust what 'low carbon' development might mean in line with existing interpretations of 'normal' urban development. From the outset, BCIL sought to provide a green gated community that entailed 'no compromise' for residents on lifestyle expectations. This in turn has meant that 'normality' in T-Zed is (re)defined with reference to existing standards in conventional housing. For example, both residents and BCIL workers have explained the problems which emerged after BCIL gave an inappropriate shiny coating to the compressed-earth blocks, which disturbed the original design and prevented the evaporation of moisture (causing the walls to bubble). In explaining why BCIL, aware that this was not the appropriate treatment for the compressed-earth blocks, consented to the coating, a resident explains:

> it is not that they didn't know ... they bowed to pressure! ... you come and say oh, this mud doesn't look nice, this is what I'm getting and how can you give me a high end house with the walls like this. ... And, he [BCIL CEO] kept responding to residents ....
> (Interview R4)

For the developers, however, the act of caring for the development has become a normal part of their operations:

> Initially we had dampness. Initially!!! Because of the rain or so. & also, the walls, the customers they have painted in their own way and then we have found bubbles in the paint, because of the paint has not been absorbed properly; then plumbing, then power problems, but normal maintenance issues, there is not anything specific for T-Zed. (Interview D4)

While the experimental nature of the development has led to the need for specific forms of maintenance, the changing social relations it has engendered have served to create an additional set of ongoing, low key, processes of repair. At the same time, entering into a social contract with residents concerning the promise of low carbon development has meant that BCIL have needed to adjust the experiment to the expectations of those who participate in it and their understanding of what 'normal' urban life might mean in Bangalore.

Furthermore, in order to maintain T-Zed as a climate change experiment, work is also required regarding the integration of the development in the neighbourhood and the city. This is a complex process, requiring negotiating about the nature of 'green' gated communities and also the reconfiguration of existing urban socio-technical systems. One area where disputes have been significant relates to land development, where legal disputes have arisen over one part of the T-Zed compound and landscaping work is at a standstill and for the four approach roads which remain dirt tracks, giving daily remainders of the development's precarious hold within the city and the tensions arising between the aspirations to provide a 'self-sufficient' development and its embedding within existing urban landscapes. The idea of a self-sufficient home, advanced here under a carbon reduction flag, contributes, on the one hand, to creating a luxury development in the context of lack of infrastructure for that type of service provision and the securitisation of resources in light of a future crisis of resources, particularly in terms of lack of water and increasing price of electricity, that is affecting the whole city. Thus, much of the discourse concerning T-Zed as a 'green' gated community focused on the idea of creating water and energy self-sufficient communities in which 'you are not taking any electricity [nor] water from the state ... you generate your own [energy and] water and then use it' (Interview C6). BCIL managers identify the limited reliance of T-Zed from urban infrastructure networks as a symbol of success:

> We are saving the cost of the entire society by lowering the load by what we are taking in electricity from the grid and by not connecting to the municipal sources ... and we should be proud to tell that we are not connected to any municipality sources for water. (Interview D2)

In this manner, T-Zed could be seen as contributing to the splintering of urbanism in Bangalore, creating enclaves within which those who can afford to secure their own resources not only through the process of making, but also through the process of maintaining. Achieving this splintered landscape, however, relies on different configurations of disconnection and reconnection with existing urban infrastructure systems. This is perhaps most obvious in the water infrastructure system, which developer and residents consider to be completely independent. For example, two residents explained how proud they were of the water service within the compound:

R3: I get the water directly from the tap and drink it ...
R2: That is a big luxury
R3: That is something that people that come to visit me takes time to understand
R2: And it also taste well, this is a big part of your life. (Interviews R2 and R3)

However, securing such water supplies has not been straightforward. Although the network of surface wells had been carefully planned by calculating the carrying capacity of the soil, the continuous droughts suffered in the city have resulted in the system not being able to recharge itself as planned and, thus, as elsewhere in the city, T-Zed has been affected by a water shortage. This in turn meant that the plans to provide water through rainwater harvesting have had to be rethought, and additional capacity has been added by building two bore-wells of significant depth which allows the development to remain independent from the municipal piped supply, but, arguably, encroaches on the collective groundwater resources of the city. Simultaneously, the water supply is affected by an alternative pattern of reconnection whereby residents in nearby informal settlements seek to use the water resource which has emerged through the development. As one of the residents in T-Zed explains:

> Apparently every morning there is a long line outside the gate, of people who want to ...
> from the village ... who want to collect water, take water from T-Zed for drinking! ... they
> don't have any drinking water! ... the [residents] association doesn't approve of it! ...
> because they don't know how many ... once you start, you really can't stop! But,
> I think the security allows a lot of people in because either they pay them or they
> empathize with them .... (Interview R4)

Despite its ambition to offer an autonomous, zero carbon, form of urban development, T-Zed is intimately connected to wider systems of water provision in the city. As this example shows, the pattern of connections and disconnections associated with maintaining low carbon experiments highlights their ambiguous character their role in reproducing existing inequalities within the city. Ecological security emerges as the central concern in developments like T-Zed:

> This [new suburb project] is a completely off-grid community. We generate our own
> power ... We had to go and dig a few wells and withdraw the water, okay, but that's a
> need out there ... [in projects like this] one is to get a more secure life ... like a secure
> way of living. In these communities, the quality of life in these communities is much
> better. Every resource available for them has been taken care of. There is enough
> water, enough power for the people that own there. (Interview D1)

While self-sufficient developments are still seen as 'a rarity in India' (Interview C6), new planning guidelines within the metropolitan area of Bangalore focus on making 'the region self-sufficient in water' (Interview C8). In this context, it is not surprising that the ability of 'green' gated communities to be self-sufficient is the major factor which has shaped their political recognition:

> the mayor and the deputy mayor ... [we tell them] if you've got these things put together
> ... your city can use much less water ... in a way that your city manages much less waste
> ... in a way that your city uses much less power or energy as demand goes. That interests
> them!
> (Interview D3)

Such claims for securing the city have to be read within a wider context. As a sub-urban development, contributing to urban sprawl, increased use of transportation and arguably contributing to maintaining unequal access to resources across an economically and socially fractured city, the question becomes one of how low carbon urbanism might serve to maintain some privileged forms of sustainability over others.

A final issue that emerges in relation to the insertion of T-Zed into the broader context of urban development is its ability to operate within the existing governance context. Several commentators have highlighted that the urban governance in Bangalore is shaped by the unequal access of upper classes to para-estatal agencies and public–private partnerships which regulate new developments (Ghosh, 2005). Initially, BCIL struggled to operate in this context: they lacked the contact and resources to deal with the planning context. Unfinished roads and gardens due to land ownership disputes are not just examples of tensions between land claims but rather serve to demonstrate that BCIL struggles with inserting the development into existing forms of governance. The ability of BCIL to secure new plots of land for new developments on the periphery of the city (in some cases encroaching on peri-urban ecosystems and livelihoods) suggests that the experience of T-Zed helped BCIL to learn and negotiate the status of the organisation within the overall system of governance of the city. The experiment in T-Zed has helped BCIL to develop effective strategies to navigate the complex map of land development in Bangalore. In maintaining the experiment, BCIL has adjusted its role to fit the existing urban development context of the city, to the extent that some of the architects that initially started with the company now regard it as being involved in 'mainstream development'.

Living low carbon

In exercising responsibilities for maintenance and repair, it is clear that experimenting with climate change is an ongoing project. A third critical dimension of this process relates to the ways in which low carbon practices become embedded within everyday life. Here, we focus on how the notion of 'carbon control' (While *et al.*, 2010) served to underpin the approach which BCIL took to the development and the ways in which it sought to shape resident's conduct in line with this logic. In promoting 'low carbon' living, BCIL developed a series of dispositions regarding everyday living in the development while simultaneously avoiding passing judgments on urban lifestyles of its customers. To this end, a series of techniques and technologies was directed towards enacting an implicit, 'material' control over these lifestyles. For example, both the air-conditioning system and the refrigerator were developed to prevent the installation of non-sustainable appliances once the house was sold. Automated control devices (called by BCIL 'conscience meters') were installed to make residents conscious of both their use of energy and its costs (environmental and economic). In so doing, the central idea is that the design of the development itself enables a certain politics of life and residence.

Combined with the logic of an alternative, no compromise, way of being low carbon and newly urban in Bangalore, these techniques for conducting daily life have had effect. The project attracted specific types of green-minded professionals who wanted to participate fully in the concept of low carbon living. Some even made T-Zed a personal project, undergoing great lengths to furnish their interiors in line with BCIL philosophy, with recycled materials or materials with low embodied energy and explicitly to focus on developing low carbon lifestyles (such as the use of electric cars). Residents highlighted the practical dimension of living low carbon that the development as engendered:

> Q: Has living in T-Zed changed your lifestyle?
> R1: Lifestyle ... lifestyle. ... Maybe our thinking has changed because before living in T-Zed we never thought about waste segregation, CFL [compact fluorescent lights] bulbs ... after we moved here we started to think environmentally ...
> R2: Unless you do it you don't realize ...
> R1: Only when you do [low carbon practices] them you realize how important the environment is ...
> [...]
> R3: I am more conscious of my carbon footprint. And about using plastics ... when I go shopping, never accept plastic bags. And [I am] being a bit more conscious, using public transportation .... (Interviews R1, R2 and R3)

Furthermore, T-Zed (and this is being reproduced in other similar gated compounds in Bangalore) is closely associated with a certain ideal of communitarian living that appears to have been transferred from the BCIL designers to T-Zed residents. As a resident explains:

> BCIL really has to be thanked for that ... because the builder initiated that kind of work or interaction and we've drawn ... we've learned to do things like that because a lot of systems do require common [management] ... there is a lot more interaction because of that ... if something is not working, you need to get together. (Interview R4)

This notion of low carbon living as a collective endeavour is enacted in a series of practices. The library, for example, collects books donated by residents. Paper is collected for recycling by a specialised NGO. Common classes for yoga and aerobics and children's activities aim to bring residents together. The garden has been designed as an amphitheatre where community events can be held. BCIL also envisaged that residents would cultivate organic food in the sky garden that they could exchange in an organic market, building a space for that purpose, although so far the residents have not engaged with this initiative. Finally residents were encouraged to form a green committee that would take responsibility for the future of the development as a whole. The centrality of establishing a low carbon community for the success of the experiment is evident in one of the resident's testimonies publicised on BCIL's website: 'The most important thing that T-Zed has been able to achieve is to get together a vibrant community of residents who make a difference in the way we live by going green in our day to day life!!' (BCIL, n.d.). This constructed community becomes the engine of the low carbon experiment, particularly in terms of determining living practices and giving the experiment a life of its own:

> Amongst the residents, there is a Green community now. ... They are conscious about the waste that they manage; they have tried to monitor certain things in terms of, you know, consumption ... in their campus. So, there is a certain amount of consciousness which I can't entirely state that we [BCIL] have instilled in them! (Interview C4)

However, installing the low carbon living ideal has not been without controversy. While BCIL has relied heavily on a design-led approach to conducting lifestyles within the development, it has also sought to regulate directly. For example, they attempted to ban the installation of bath tubs in individual properties, as it was considered that this created an unsustainable demand of water and disrupted the collective design of T-Zed:

> there were issues of installing bathtubs. For us, we couldn't make any sense of it, because for us a bathtub is a 250 litre butt. ... I mean, at the minimum. Whereas, a bucket butt could be 15–20 litres and a shower is about 25–30 litres, if you are a little prudent in your usage. So, we could not encourage ... so if you see, part of the dossier [given to residents] also said bathtubs are not allowed. (Interview C4)

This, however, has caused controversy among local residents, as one of them explained:

> I think they got a bit more of the stick! ... for example ... oh, that the houses should not have bathtubs! ... now, in India people don't use bathtubs anywhere! ... people do buckets of water or they do showers, right! But, I think it is just as natural that people like us, where we are getting to a certain level of life, we just wanted to have a bathtub in the house! I don't think you would end up using it more than once or twice a year ... but we would just like to have the option! ... Some of the people in the organization really got on this high and moral horse ... they spent an inordinate amount of energy on things like that, on a discussion ... they got very emotional and a very moralistic kind of thing! (Interview R5)

Direct forms of intervention such as this have been met with hostility. Having already put themselves forward as responsible citizens by buying into the zero carbon ideal, some local residents perceive that additional requirements to conduct their lives in a low carbon manner are unrealistic and unfair. However, residents appear less concerned by the potential regulatory attempts within their own community (even though this is also the result of BCIL efforts) which is regarded as a legitimate means through which collective decisions concerning lifestyles might be met. Despite the implicit acceptance of the low carbon design of the development and the requirements it makes in terms of everyday practices, residents did also express the sense that living with technology' became a burden, in the sense that people living in their own homes did not want to be continuously worried about the technologies they dwell with and whether or not they work. During the construction, for example, improvised solutions had to be found to fit too many types of materials which did not behave exactly as designers had predicted. The highly customised fridges and air-conditioning systems are sources of frustration for developers, consultants and residents. While failure may be an inherent risk within the process of experimentation, residents have started to question whether the design team went too far in enacting innovation. Indeed, BCIL is now developing new projects with fewer innovations in order to reach a higher market and establish itself as a credible builder—even though this move has been described by critics as 'going back to conventional development' (Interview D1).

Conclusions

While conventional accounts of urban responses to the climate change problematic have focused on the processes through which knowledge, institutions and capacity are built within cities and the processes of designing and implementing policy, this paper has suggested that this provides only a partial account of the dynamics at work. Focusing on the ad-hoc arrangements, initiatives, projects and interventions that can be termed 'climate change experiments', it has been argued that they constitute an important means through which governing climate change in the city is taking place. In order to understand the processes of climate change experimentation and their potential within wider urban contexts, a threefold analytical framework has been advanced that examines making, maintaining and living low carbon as three inter-related processes. In so doing, attention is drawn to the socio-technical processes involved in creating climate change experiments, the ongoing work involved in maintaining the experiment within existing urban networks, and the importance of aligning the experiment within the context of everyday life. In this manner, climate change experiments become critical means through which understanding what it might mean to be low carbon, technically, socially, economically and politically, takes place and, through insertion within wider urban socio-technical networks, carry the potential to reconfigure processes of urban development and the conduct of daily life.

In the case of T-Zed, it was found that the processes of making, maintaining and living low carbon are contributing to an emergent low carbon urbanism in three principal ways. First, T-Zed has created a space for climate change innovation. Whereas, as explained above, the techniques used in T-Zed have been applied in other contexts around the city, T-Zed is the first development which incorporates both vernacular ideas about housing and Western appliances at this scale. As a result of the success of the project in these terms, BCIL are involved with other urban developments in the city and currently provide advice to the Indian government on the potential for an energy conservation building code, with potentially national scale application. Apart from replicating T-Zed experience in new BCIL projects, it has provided the space for new companies to develop their products and concept and move on to wider markets (e.g. Flexitron, a light innovation company) and seeded new consultancy companies working in the field. Further, the experiment has provided a space for innovation concerning what urban development in Bangalore could and should entail, for example as is being taken forward in the work of one new developer, Samskruti Builders, whose leaders recognise the influence of T-Zed is seeing the 'green idea' realised for the first time. This influence can be seen in Samskruti projects, for example, in their focus on local materials and vernacular architecture, their interest of creating self-sufficient communities. In making a low carbon experiment, BCIL has created the space within which thinking of development as low carbon has become possible in the Bangalore context. This has not taken place through establishing a protected niche for new forms of technology or social innovation, but rather than the complex dynamic of making, maintaining and living low carbon development through which T-Zed is variously connected into, and separated from, wider urban political ecologies.

Second, the presence of T-Zed within the urban landscape of Bangalore is serving to transform and reconfigure urban infrastructure systems more broadly. As discussed above, maintaining the experiment relies on simultaneous processes of disconnection—of the archetypal 'gated' community—and of reconnection—through urban land disputes, to the water table, and through relations with neighbouring informal settlements and other gated communities. While the self-sufficiency of T-Zed is regarded as a key element in its quest towards zero carbon development, it is also a means through which such developments can create markets by promising 'security' of energy and water supplies. Despite intentions to provide independent water supplies, this particular case relies on borehole wells, creating a deep connection to the city's future water security. At the same time, the continued discourse of self-reliance may serve to work against arguments for the extension of universal provision and access to water and energy supplies. While issues of ecological security may provide a 'Trojan horse' through which low carbon urbanism can enter into the wider urban infrastructure networks across the city, they may just as well serve to reproduce existing inequalities in terms of access to resources and to sustainable livelihoods. This in turn suggests that it is important to open up such forms of low carbon urbanism to critical scrutiny, examining who benefits and who loses from emerging discourses of urban ecological security, carbon control and resilience that are being advanced in the name of responding to climate change.

Third, T-Zed has provided an arena in which new discourses of responsibility and carbon control for middle class residents have flourished. By adopting an approach which suggests that low carbon living is compatible with modern urbanism in Bangalore, the development provides a 'showcase' for other middle-class communities. Indeed, considerable effort goes into visible symbolic acts which ratify the green status of residents, and residents themselves have been involved in activities to spread the message about the potential of 'green' living virally. However, this is a process which is highly contested as, on the one hand, residents accept forms of co-regulation from their peers but challenge the direct intervention of BCIL, as a developer, in their everyday lives. The ambivalence of this position is quite clear. Under the motto of no compromise, BCIL has provided spacious designed apartments with luxuries such as air-conditioning, swimming pool and squash court. Yet, they have attempted to regulate the use of water and the cultivation of organic vegetables in the terraced gardens. When residents protest about BCIL jumping 'on the high moral horse' they are questioning BCIL's capacity and legitimacy to demarcate which practices are sustainable and which are not.

In seeking to normalise climate change, by fostering more mainstream forms of low carbon development, reconfiguring existing urban networks, and placing carbon control as part of everyday life, the T-Zed project demonstrates the ways in which experiments play a role in governing climate change in the city. The results here, as elsewhere, are deeply ambivalent, serving both to challenge and to sustain existing forms of urban development and inequality. While this kind of green development constitutes an alternative to conventional forms of gated community for a growing market of high-earning green-minded middle-class residents, it falls some distance from solving the fundamental problems of inequality and infrastructure provision in Bangalore.

Acknowledgements

The research upon which this paper is based was supported by Harriet Bulkeley's ESRC Climate Change Fellowship *Urban Transitions: climate change, global cities and the transformation of socio-technical networks* (Award Number: RES-066-27-0002). We are also grateful to the participants in the research for giving their time and insights to the project, and to the reviewers of this paper for their useful comments which have helped us to improve our arguments.

Notes

1. 'Experiment' is a complex word, conveying several different meanings, including the familiar scientific sense of 'testing' a hypothesis under controlled conditions, but also, and more importantly for present purposes, it can mean a 'tentative procedure' and the 'action of trying anything, putting it to proof', as well as 'to have experience of … to feel' (OED Online). To be experimental is about the process of experience, of bearing 'witness' (OED Online) and is a tentative, unfolding, process, rather necessarily implying something novel or innovative.
2. This case was conducted as one of five studies of 'climate change experiments' undertaken as part of the ESRC Climate Change Fellowship, Urban Transitions: climate change, global cities and the transformation of socio-technical networks (Award Number: RES-066-27-0002).
3. Although in some cases obtaining government clearances has delayed some sustainability projects, particularly if they are on a large scale, the control of government over developers is loose, particularly when developers are not concerned with integrating a particular community within the city's energy, water or waste infrastructure.

References

Alber, G. & Kern, K. (2008) Governing climate change in cities: modes of urban climate governance in multi-level systems, *Proceedings of the OECD Conference on Competitive Cities and Climate Change* (Paris, Organisation for Economic Co-operation and Development (OECD)).

Bai, X. (2007) Integrating global environmental concerns into urban management: the scale and readiness arguments, *Journal of Industrial Ecology*, 11(2), 15–29.

Betsill, M. & Bulkeley, H. (2007) Looking back and thinking ahead: a decade of cities and climate change research, *Local Environment: International Journal of Justice and Sustainability*, 12(5), 447–456.

Biodiversity Conservation India Limited (BCIL) (2009) *UNEP—case study* (Bangalore, BCIL).

Biodiversity Conservation India Limited (BCIL) (n.d) *T-Zed testimonials*, Available online at: http://ecobcil.com/content/bcil-t-zed?livserv_cs_id=6458072 (accessed 20 December 2010).

Bulkeley, H. (2010) Cities and the governing of climate change, *Annual Review of Environment and Resources*, 35, 229–253.

Bulkeley, H. & Kern, K. (2006) Local government and the governing of climate change in Germany and the UK, *Urban Studies*, 43(12), 2237–2259.

Bulkeley, H., Castán Broto, V., Hodson, M. & Marvin, S. (Eds.) (2011) *Cities and low carbon transition* (Abingdon, Routledge).

CIVIC (2008) *Pilot slum and urban homelessness study* (Bengaluru, Bruhath Bengaluru Mahanagara Palike (BBMP)).

Corfee-Morlot, J., Kamal-Chaoui, L., Donovan, M. G., Cochran, I., Robert, A. & Teasdale, P. J. (2009) *Cities, climate change and multilevel governance*. OECD Environmental Working Papers 14 (Paris, OECD Publishing).

Dhakal, S. (2009) Urban energy use and carbon emissions from cities in China and policy implications. *Energy Policy*, 37(11), 4208–4219.

Dodman, D. (2009) Blaming cities for climate change? An analysis of urban greenhouse gas emissions inventories, *Environment and Urbanization*, 21(1), 185–201.

Geels, F. W. (2002) Technological transitions as evolutionary reconfiguration processes: a multi-level perspective and a case-study. *Research Policy*, 31(8–9), 1257–1274.

Ghosh, A. (2005) Public-private or a private public? Promised partnership of the Bangalore agenda task force. *Economic and Political Weekly*, 40(47), 4921–4922.

Gore, C. & Robinson, P. (2009) Local government response to climate change: our last, best hope? in: H. Selin & S. D. VanDeveer (Eds) *Changing climates in North American politics: institutions, policymaking and multilevel governance* (Cambridge, MA, MIT Press), 138–158.

Graham, S. & Marvin, S. (2001) *Splintering urbanism: networked infrastructures, technological mobilities and the urban condition* (London, Routledge).

Granberg, M. & Elander, I. (2007) Local governance and climate change: reflections on the Swedish experience, *Local Environment: International Journal of Justice and Sustainability*, 12(5), 537–548.

Hajer, M. (2003) Policy without polity? Policy analysis and the institutional void, *Policy Sciences*, 36, 175–195.

Hodson, M. & Marvin, S. (2009) 'Urban ecological security': a new urban paradigm? *International Journal of Urban and Regional Research*, 33(1), 193–215.

Hoffman, M. (2009) Experimenting with climate governance, paper presented at the *2009 Amsterdam Conference on the Human Dimensions of Global Environmental Change—Earth System Governance: People, Places and the Planet*, Amsterdam, the Netherlands, December.

Hoffman, M. J. (2011) *Climate governance at the crossroads: experimenting with a global response* (New York, Oxford University Press).

Hoornweg, D., Sugar, L. & Gomez, C. L. T. (2011) Cities and greenhouse gas emissions: moving forward, *Environment and Urbanization*, 23(1), 207–227.

International Energy Agency (IEA) (2008) *World energy outlook 2008* (Paris, International Energy Agency).

Jollands, N. (2008) Cities and energy—a discussion paper, paper presented at the *OECD International Conference Competitive Cities and Climate Change*, Milan, Italy.

Keil, R. (2005) Progress report—urban political ecology, *Urban Geography*, 26(7), 640–651.

Keivani, R. & Mattingly, M. (2007) The interface of globalization and peripheral land in the cities of the south: implications for urban governance and local economic development, *International Journal of Urban and Regional Research*, 31(2), 459–474.

Monstadt, J. (2009) Conceptualizing the political ecology of urban infrastructures: insights from technology and urban studies, *Environment and Planning A*, 41(8), 1924–1942.

Puppim de Oliveira, J. A. (2009) The implementation of climate change related policies at the subnational level: an analysis of three countries, *Habitat International*, 33(3), 253–259.

Ranganathan, M., Kamath, L. & Baindur, V. (2009) Piped water supply to Greater Bangalore: putting the cart before the horse? *Economic and Political Weekly*, 44(33), 53–62.

Revi, A. (2008) Climate change risk: an adaptation and mitigation agenda for Indian cities, *Environment and Urbanization*, 20(1), 207–229.

Romero Lankao, P. (2007) How do local governments in Mexico City manage global warming? *Local Environment: International Journal of Justice and Sustainability*, 12(5), 519–535.

Romero Lankao, P. (2010) Water in Mexico City: what will climate change bring to its history of water-related hazards and vulnerabilities? *Environment and Urbanization*, 22(1), 157–178.

Rutland, T. & Aylett, A. (2008) The work of policy: actor networks, governmentality, and local action on climate change in Portland, Oregon, *Environment and Planning D: Society and Space*, 26(4), 627–646.

Satterthwaite, D. (2008) Cities' contribution to global warming: notes on the allocation of greenhouse gas emissions, *Environment and Urbanization*, 20(2), 539–549.

Satterthwaite, D. (2011) Editorial: Why is community action needed for disaster risk reduction and climate change adaptation? *Environment and Urbanization*, 23(2), 339–349.

Schreurs, M. A. (2008) From the bottom up: local and subnational climate change politics, *Journal of Environment Development*, 17(4), 343–355.

Smith, A. (2007) Translating sustainabilities between green niches and socio-technical regimes, *Technology Analysis and Strategic Management*, 19, 427–450.

Smith, A., Voss, J. & Grin, J. (2010) Innovation studies and sustainability transitions: the allure of the multi-level perspective and its challenges, *Research Policy*, 39, 435–448.

Sugiyama, N. & Takeuchi, T. (2008) Local policies for climate change in Japan, *Journal of Environment Development*, 17(4), 424–441.

While, A., Jonas, A. E. G. & Gibbs, D. C. (2010) From sustainable development to carbon control: eco-state restructuring and the politics of urban and regional development, *Transactions of the Institute of British Geographers*, 35, 76–93.

World Bank (2010) *World Development Report 2010: Climate change* (Washington, DC, The World Bank).

Index

INDEX